RENEWALS 458-4574

Mastering the Acquirer's Innovation Dilemma

Mastering the Acquirer's Innovation Dilemma

Knowledge Sourcing through Corporate Acquisitions

Valerie Bannert-Thurner

 in association with the European Institute for Technology and Innovation Management

© Valerie Bannert-Thurner 2005

All rights reserved. No reproduction, copy or transmission of this publication may be made without written permission.

No paragraph of this publication may be reproduced, copied or transmitted save with written permission or in accordance with the provisions of the Copyright, Designs and Patents Act 1988, or under the terms of any licence permitting limited copying issued by the Copyright Licensing Agency, 90 Tottenham Court Road, London W1T 4LP.

Any person who does any unauthorized act in relation to this publication may be liable to criminal prosecution and civil claims for damages.

The author has asserted her right to be identified as the author of this work in accordance with the Copyright, Designs and Patents Act 1988.

First published in 2005 by
PALGRAVE MACMILLAN
Houndmills, Basingstoke, Hampshire RG21 6XS and
175 Fifth Avenue, New York, N.Y. 10010
Companies and representatives throughout the world.

PALGRAVE MACMILLAN is the global academic imprint of the Palgrave Macmillan division of St. Martin's Press, LLC and of Palgrave Macmillan Ltd. Macmillan® is a registered trademark in the United States, United Kingdom and other countries. Palgrave is a registered trademark in the European Union and other countries.

ISBN-13: 978–1–4039–4755–0
ISBN-10: 1–4039–4755–4

This book is printed on paper suitable for recycling and made from fully managed and sustained forest sources.

A catalogue record for this book is available from the British Library.

Library of Congress Cataloging-in-Publication Data
Bannert-Thurner, Valerie, 1978–
    Mastering the acquirer's innovation dilemma : knowledge sourcing through corporate acquisitions / by Valerie Bannert-Thurner.
    p. cm.
    Includes bibliographical references and index.
    ISBN 1–4039–4755–4 (cloth)
    1. Consolidation and merger of corporations. 2. Technological innovations.
3. Strategic planning. I. Title.
HD2746.5.B34 2005
658.1′62—dc22   2005047468

10 9 8 7 6 5 4 3 2 1
14 13 12 11 10 09 08 07 06 05

Printed and bound in Great Britain by
Antony Rowe Ltd, Chippenham and Eastbourne

To Philipp

Contents

List of Figures — xi

List of Best Practices — xv

List of Abbreviations — xvi

Preface — xix

1 Introduction — 1

   1.1 Relevance of the Topic – Acquisitions in Innovation-driven Industries — 1
   1.2 Acquisition Performance from a Technology-based Perspective — 5
   1.3 Call from Practice – Mastering the Innovation Dilemma — 7
   1.4 Objectives, Research Approach and Content of the Book — 9

2 Technology, Innovation and their Management — 13

   2.1 Terms and Definitions: Technology-based Value Creation — 13
   2.2 Technology Management — 17
        2.2.1 Integrated technology management — 17
        2.2.2 Basic elements of the integrated technology management — 18
        2.2.3 Main tasks of the integrated technology management — 23
   2.3 Innovation Management — 26
   2.4 Conclusion — 27

3 Introduction to Corporate Acquisitions — 28

   3.1 Terms and Definitions: Corporate Acquisitions — 28
   3.2 Acquisition Types — 29
   3.3 Acquisitions as a Means of Value Creation — 30
   3.4 Acquisitions as a Business Strategic Path — 30
   3.5 Conclusion — 31

4 Case Studies from Reality: Technology-based Value Creation in Real-life Acquisitions — 32

    4.1 Theoretical Model to Investigate Technology-based Value Creation — 33
    4.2 Selection of the Case Studies and Data Collection — 34
    4.3 Cases from Reality — 37
        4.3.1 Case 1: Hilti–Ammann Lasertechnik acquisition — 37
        4.3.2 Case 2: Fillpack–Aseptofill acquisition — 46
        4.3.3 Case 3: Phonak–Unitron acquisition — 57
        4.3.4 Case 4: Unaxis–Plasma-Therm acquisition — 69
        4.3.5 Case 5: Starrag–Heckert acquisition — 86
    4.4 Conclusion on Cases from Reality — 99

5 Model of Reality: A New Understanding of Technology-based Value Creation in Corporate Acquisitions — 100

    5.1 Technology-based Value Creation – Quick gains and Long-term Success — 101
        5.1.1 Innovation and resource deployment — 101
        5.1.2 Quick gains and long-term successes — 102
    5.2 Acquisition Type – Indeed not all M&As are Alike – and that matters — 104
        5.2.1 Acquisition characteristics — 105
        5.2.2 Acquisition types and strategies — 110
    5.3 Initial Conditions – Necessary, However Insufficient, Conditions — 113
        5.3.1 Technology-based value creation potential — 113
        5.3.2 Acquisition capability — 119
        5.3.3 Strategic fit — 121
        5.3.4 Contextual fit — 123
    5.4 Strategy Processes – The Appropriate Acquisition and Integration Processes — 125
        5.4.1 The impact of the acquisition process on technology-based value creation — 126
        5.4.2 The impact of the integration process on technology-based value creation — 129
    5.5 External Developments — 135
    5.6 Conclusion from the New Understanding — 136

6 Technology-based Strategic Acquisition and Integration Management — 137

- 6.1 Developing the Concept — 138
- 6.2 Overview of the Concept — 139
  - 6.2.1 Phases of the technology-based acquisition and integration management — 139
  - 6.2.2 Layers of the technology-based acquisition and integration management — 141
  - 6.2.3 Technology-based strategic acquisition and integration management on the normative, strategic and operational level — 145
  - 6.2.4 Integrating the technology and innovation management into the AIM — 145
- 6.3 The Detailed Concept — 145
  - 6.3.1 Strategic planning — 148
  - 6.3.2 Company screening — 154
  - 6.3.3 Acquisition and integration strategy development — 166
  - 6.3.4 Due diligence — 202
  - 6.3.5 Integration planning — 220
  - 6.3.6 Strategy implementation — 228
- 6.4 Conclusion from Technology-based Strategic Acquisition and Integration Management — 234

7 Management Principles — 237

- 7.1 Management Principles for a New Understanding — 237
  - 7.1.1 Management Principle 1: Wheel of technology-based value creation — 237
  - 7.1.2 Management Principle 2: Innovativeness cannot be bought — 238
  - 7.1.3 Management Principle 3: Willingness to change as key initial condition — 238
  - 7.1.4 Management Principle 4: Creative destruction of innovation context — 239
- 7.2 Management Principles for a New Concept — 239
  - 7.2.1 Management Principle 5: Technology concerns as basic fundamental — 239
  - 7.2.2 Management Principle 6: Holistic, systemic and integrative concept — 240
  - 7.2.3 Management Principle 7: Prospective perspective — 240
- 7.3 Management Principles for Implementation — 241
  - 7.3.1 Management Principle 8: Implementation requires a new awareness of acquisitions — 241

  7.3.2 Management Principle 9: Implementation requires
    a new awareness of technology and innovation 241
  7.3.3 Management Principle 10: Implementation
    requires critical self reflection 242

Appendix A – Value Creation in Corporate Acquisitions 243

Appendix B – Checklist: Technology Due Diligence 251

Notes 257

Bibliography 267

Index 281

List of Figures

1.1	Five merger waves in the USA	2
1.2	Changes increase the importance of a technology-based perspective on acquisitions	4
1.3	Overview on acquisition performance from a technology-based perspective	6
1.4	Research design of this book	11
1.5	Outline of this book	12
2.1	Five resource integration mechanisms leading to innovation and efficient resource deployment	16
2.2	'Integrated Technology Management'	19
2.3	Functional match between technologies and needs	20
2.4	Innovation architecture	21
2.5	Trilogy of strategic technology decisions	22
2.6	Technology strategy formulation and implementation and associated tasks	23
2.7	Technology assessment methods depending on degree of uncertainty	25
3.1	Corporate acquisitions	29
4.1	Influential aspects on the successful realization of value creation	33
4.2	Overview of case studies	38
4.3	Acquisition timeline of Hilti–Ammann Lasertechnik acquisition	39
4.4	Potential for a new business	40
4.5	Rotating laser from Ammann Lasertechnik used for positioning and alignments	42
4.6	Distant organizational integration of Amman Lasertechnik	43
4.7	Time-to-market process at Hilti	44
4.8	Acquisition timeline of Fillpack–Aseptofill acquisition	47
4.9	Strategic competencies in the value chain of aseptic filling of plastic bottles	48
4.10	Aseptofill activities match the strategic focus of Fillpack	50
4.11	Value-creation opportunities pursued through the Aseptofill acquisition	52
4.12	Integration teams at the Aseptofill acquisition	53
4.13	Trouble shooting and technological synergies at the Aseptofill filling machine	54

4.14	Joint development projects to transfer technologies and develop new products	55
4.15	Fillpack Holding's press release, 28 January 2003	56
4.16	Acquisition timeline of Phonak's Unitron acquisition	59
4.17	Acquisition team at the Unitron acquisition	61
4.18	Initial organizational integration of Unitron	63
4.19	Product concept of the Aero	64
4.20	Product configuration of the Unison	66
4.21	Acquisition timeline of Unaxis's Plasma-Therm acquisition	70
4.22	Unaxis's organizational structure in 1999	71
4.23	Rough innovation architecture of the SC unit and Leybold's magneto business	72
4.24	Thin film-related competencies within the semiconductor value chain	74
4.25	Acquisition process of the Plasma-Therm acquisition	76
4.26	Value-creation opportunities in the Plasma-Therm acquisition	78
4.27	Integration teams of the Plasma-Therm acquisition	79
4.28	Modular structure of a Unified Platform product	82
4.29	Rough roadmap of the Unified Platform	84
4.30	Acquisition timeline of Starrag's Heckert acquisition	87
4.31	External developments within the Starrag markets	88
4.32	Fit of product portfolios between Starrag and Heckert	89
4.33	Expected synergies from acquiring Heckert	90
4.34	Organizational integration of Heckert with boundary-spanning structures	92
4.35	SX-051, Platform 3 developed on Heckert's base machine	93
4.36	The Allegra project as the future product pipeline	94
4.37	The platforms in all sizes will serve Starrag and Heckert markets	95
4.38	Modular building blocks for Starrag and Heckert product platforms	96
4.39	Product integration roadmap: platform strategy	97
5.1	Elements of the model of understanding	101
5.2	Planned and emergent technology-based value-creation opportunities	102
5.3	Wheel of technology-based value creation	103
5.4	Characteristics of the acquisition types	106
5.5	Different resource-integration mechanisms for different business relations	107
5.6	Organizational and contextual integration for relatively small and large acquisitions	108
5.7	Technology development and related curves	109

5.8	Characteristics of the three acquisition types: Ⓐ venturing acquisitions, Ⓑ substrate-for-growth acquisitions, Ⓒ play-for-scale acquisitions	110
5.9	Summary of the acquisition types and characteristics	111
5.10	Necessary, however insufficient, conditions	113
5.11	Technology-based value-creation potential	114
5.12	Attributes of the resources to be combined and associated value-creation potential	116
5.13	Resource relation and associated technology-based value creation	118
5.14	Strategic relevance of acquisitions and other transaction forms	122
5.15	Contextual fit dependent on contextual characteristics and similarities	123
5.16	Impact of contextual similarities on technological synergies	124
5.17	Model of the acquisition integration process	132
6.1	Technology-based strategic acquisition and integration management	140
6.2	Different layers of the technology-based acquisition and integration process	143
6.3	A&I process on the normative, strategic and operational levels	146
6.4	Integrating strategic, technology, and innovation, and acquisition and integration management	147
6.5	Integrated strategic planning	149
6.6	Planning cycles to allow strategic acquisition	150
6.7	Strategic acquisition options which extend scale and scope of the company	151
6.8	Self due diligence	153
6.9	Screening process	156
6.10	Source of information	158
6.11	Requirements for achieving successful value creation	159
6.12	Evaluation of potential target companies	160
6.13	Strategic fit in different functional areas	161
6.14	Core competence check	162
6.15	Rough contextual fit evaluation	163
6.16	Development of the acquisition and integration strategy	168
6.17	Objectives, elements and methods of a technology and innovation analysis	169
6.18	Analysis of the target's resources	171
6.19	Attributes of the target's corporate context	172
6.20	Merging and integrating the target's and acquirer's resource bases	175
6.21	Identification of technology-based value-creation opportunities	177

6.22	Market-pull and technology-push technology-based value-creation opportunities	179
6.23	Value chain analysis for the identification of resource deployment synergies	180
6.24	Optimization of technology deployment	182
6.25	Documentation of value-creation opportunity	183
6.26	Evaluation of value-creation opportunities	184
6.27	Three order fit paths	185
6.28	Net present value calculation	186
6.29	VCO portfolio and norm strategies	188
6.30	Timeliness of value-creation opportunity	189
6.31	Developing integration strategy	191
6.32	Integration process-related acquistion strategy, integration approach and plan	192
6.33	Factors determining the integration approach	194
6.34	Example of an acquisition structure	200
6.35	Profile description of a transition manager	201
6.36	Due diligence	203
6.37	Technology due diligence	204
6.38	Three levels of the technology due diligence	204
6.39	Process of the first-level technology due diligence	205
6.40	Dynamic technology portfolio	208
6.41	Required level of combinability depending on required VCO	209
6.42	Process of the second-level technology due diligence	211
6.43	Assessment of congruence and relatedness using innovation architecture	212
6.44	Process of the third-level technology due diligence	214
6.45	Contextual web of the target's and acquirer's characteristics	216
6.46	Calculation of contextual fit	217
6.47	Example of the calculation of contextual fit	218
6.48	Integration planning process	221
6.49	Three layers of integration activities	223
6.50	Scheduling activities within different integration phases	225
6.51	Integration schedule within product development	226
6.52	Three views of the integration roadmap	227
6.53	Strategy implementing	231
6.54	Integration performance measurement	232
6.55	Example of an integration structure	233
6.56	Integration policy	234
6.57	Summary of key issues necessary to achieve technology-based value creation	235
A.1	Literature overview of aspects relevant to value creation in corporate acquisitions	248

List of Best Practices

1	Acquisition and integration management at GE	142
2	Acquisition and integration management at IBM	144
3	Strategic planning at Phoenix	152
4	Company screening at Cisco	164
5	Understanding of merging resources at Cisco	176
6	Structural integration measures at Intel	195
7	Integration management at Cisco	196
8	Technology-based value creation fostering company configuration at CIBA	197
9	Innovation fostering context on the normative level at Hewlett Packard	199
10	Transition managers at IBM	201
11	M&A Challenging Committee at CIBA	202
12	Intellectual property due diligence at Unaxis	207
13	Integration steering, controlling and learning at Siemens	229
14	Integration process at ABB	230

List of Abbreviations

A&I	acquisition and integration
ABB	Asea Brown Boveri
Admin	administration
AG	Aktiengesellschaft
AIM	acquisition and integration management
ASIC	Application-Specific Integrated Circuit
BASF	Badische Anilin- und Soda-Fabrik AG
BD	business development
bn	billion
BU	business unit
CAD	computer aided design
COO	chief operating officer
CVD	chemical vapour deposition
EBITA	earnings before interest, taxes and amortization
ERP	enterprise resource planning
FPY	First Pass Yield
GE	General Electric
GmbH	Gesellschaft mit beschränkter Haftung
HDP	high density plasma
HDPE	high density polyethylene
HP	Hewlett Packard
HR	human resources
IBM	International Business Machines
ICP	inductively coupled plasma
Innov.	innovation
Invest.	investment
IP	intellectual property
IPO	initial public offer
IT	information technology
LBO	leveraged buy-out
Lit.	literature
LT	Lasertechnik
m	million
M&A	mergers and acquisitions
Manuf.	manufacturing
MBO	management buy-out
MEMS	micro-electronic mechanical systems
Mgmt	Management

Mgr	Manager
NA	North America
NIH	not invented here
NPV	net present value
O&L	operations and logistics
OBH	Oerlion Bührle Group
Opport.	opportunity
PECVD	plasma-enhanced chemical vapour deposition
PET	polyethylene terephthalate
PF	platform
PMI	post-merger integration
Prod.	production
PVD	physical vapour deposition
R&D	research and development
RC	research centre
RIE	reactive ion etching
ROI	return on investment
RQ	research question
SBU	strategic business unit
SC	semiconductor
SC	steering committee
SiGe	silicium germanium
Str.	strategic
SWOT	strengths, weaknesses, opportunities, threats
T&I	technology and innovation
TIM	technology and innovation management
TM	technology management
TRIZ	Theory of Inventive Problem Solving
TTM	time-to-market
US	United States
USA	United States of America
VC	venture capital
VCO	value creation opportunity
VD	value driver
vs.	versus
ww	worldwide

Preface

This book is the result of my research activities at the Swiss Federal Institute of Technology, ETH Zurich, Technology and Innovation Management Group chaired by Professor Dr Hugo Tschirky. It is targeted towards professionals in the fields of acquisition and integration management and technology and innovation management as well as to academics researching those fields. The book's main objective is to outline the acquirer's innovation dilemma, to explain the key success factors to achieve long-term innovativeness after acquisitions and to introduce management concepts to master technology-intensive acquisitions. The book's contribution to the state of the art in theory and practice would certainly not have been possible without the great support from people in academia and practice, in professional and private life.

I am greatly indebted to my PhD thesis advisor, Professor Dr Hugo Tschirky from the ETH Zurich. For his professional input, his dedicated time and his general support for my research and industry projects which greatly influenced the quality of my book in a very positive way. Furthermore, he provided me with freedom and support for innovative ideas and created the excellent atmosphere in our team which made my research activities a great pleasure.

Many thanks go to Professor Markus Meier from ETH Zurich for co-advising my PhD thesis. Collaboration was a great pleasure and his challenging remarks, questions and support for my research and industry projects influenced my book very positively.

I would also like to thank my friends and research colleagues at the ETH Technology and Innovation Management Group: Gaston Trauffler, Tim Sauber, Jean-Philippe Escher, Martin Luggen, Dr Stefan Koruna and Philip Bucher. This team – my dear friends – was the reason that at no time during my research did I fail to enjoy working on my PhD. The fact that this book appears in English is due to the merit of my wonderful proofreader, Hilda Fritze-Vomvoris, who supported me greatly in achieving all of my urgent deadlines. Many thanks go also to numerous students in our department and to junior assistants in our team.

The laboratory of this research was industry. Thus I would like to express my gratitude to my many research and interview partners. Without their openness to addressing highly sensitive topics and their remarks on my concepts, this book could not make any useful contributions.

Last but certainly not least, I would like to express my gratitude to my beloved ones – my parents and my husband Philipp Thurner. Without your understanding, support and particularly your encouragement I would never have been able to finalize my doctoral thesis and finally this book. This book is dedicated to you!

Santa Barbara VALERIE BANNERT-THURNER

1
Introduction

In the recent past corporate acquisitions have drastically changed the business landscape[1] in innovation-driven industries.[2] An increasing number of acquisitions within these industries can be observed and, in particular, acquisitions where the main objective is to obtain technological competencies and to foster innovativeness have become more popular.[3] Furthermore, the acquisition's effect on the company's innovativeness and underlying resource base has become a crucial factor impacting on competitiveness and sustained profitable growth. Nevertheless, acquisitions nowadays mostly result in a lowered innovation rate (Hitt *et al.*, 1991b) and managers rarely achieve successful internalization or the efficient deployment of the acquired competencies (Capron & Mitchell, 1998). This is partly due to the limited approach researchers and practitioners apply to the topic of corporate acquisitions and their integration by addressing solely the financial, legal and economic aspects as separate concerns rather than taking a more holistic approach (Larsson & Finkelstein, 1999) which would also incorporate a technology-based perspective. Thus there is a lack of understanding of how to achieve increased innovativeness and an efficiently deployed resource base in corporate acquisitions and of applicable concepts of the strategic acquisition and integration management to foster this technology-based value creation. Triggered by these observations, this book aims at increasing the understanding of corporate acquisitions from a technology-based perspective and at proposing applicable concepts for practice which support managers in increasing their innovativeness and in efficiently deploying their competencies through corporate acquisitions.

1.1 Relevance of the Topic – Acquisitions in Innovation-driven Industries

Corporate acquisitions are an important strategic issue which has attracted much discussion and consideration in theory and practice. Generally, acquisitions foster the growth of a firm (Penrose, 1959: 155); however, the

2 Mastering the Acquirer's Innovation Dilemma

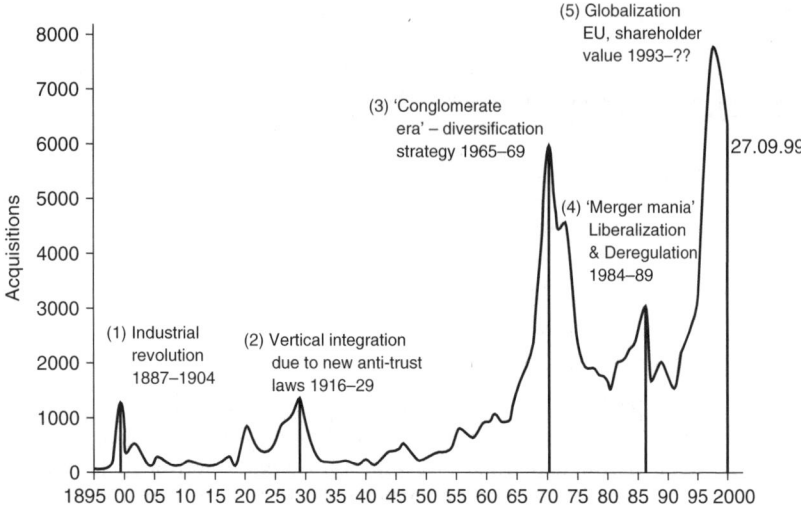

Figure 1.1 Five merger waves in the USA
Source: adapted from Müller-Stewens (2000: 44).

underlying specific motives and strategic rational are diverse and have changed over time (see Figure 1.1).

Despite the general drop in acquisition activity over the last few years and only a slight recent increase (Mergerstat, 2004), a new motive guiding acquisitions has become more and more apparent. An increasing number of corporate acquisitions were driven by the primary focus on acquiring knowledge (Davenport & Prusak, 1998: 117), fostering innovativeness, internalizing competencies and IP,[4] etc. For example companies such as Intel, Microsoft, Sun, Oracle, Genentech, Amgen and Genzyme have built their industry leading innovativeness on the ideas of others in the recent past. Also the upcoming restructuring phase within the IT Industry (Chappuis et al., 2004) and the increasing technology convergence (PriceWaterhouseCoopers, 2003) cause an increase in technology intensive and innovation-driven acquisitions. These trends are also supported by the general increase in acquisitions in high-tech industries (OECD-Economic-Outlook, 2003).[5] In addition, according to Bloomberg 342 acquisitions were registered alone in the software industry in the first quarter of 2004, which is the highest acquisition density since 2000 with a volume of $5.2 bn for the 163 published transactions.

These increasingly important innovation-driven acquisitions can be characterized as strategic acquisitions with the primary objective to achieve technology-based value creation. Technology-based value creation[6] is defined as *the short- and especially long-term value creation*[7] *derived from innovations and the efficient deployment of resources*. Whereas *innovation* is referred to as the *successful initial commercialization of the integration of resources such as knowledge, technologies, etc. into new products, resource deployment* is concerned with the *efficient transfer or substitution of resources*. Value creation from resource deployment derives from cost cutting and from building the basis for future innovations.

These innovation-driven corporate acquisitions mainly occurring in innovation-driven industries have increasingly been *driven by two upcoming forces*. These two forces are the new sources of competitiveness in innovation-driven industries. They are brought about by the current changes in scientific and technological advances, social values and industrial dynamics[8] (see Figure 1.2).

The first new driving force for innovation-driven acquisitions is the increasing need to innovate. Technological trends such as an increasing technological complexity (Tschirky, 1998a), an explosion of knowledge creation and shortening product life cycles (Tschirky, 1998a) against the background of globalization (Porter, 1986) and hyper-competition (D'Aveni, 1994) are increasingly challenging companies to raise their level of innovativeness. Tushman and Anderson (1997) put it briefly while referring to the economist Josef Schumpeter: 'the prime driver of economic progress is technological innovation' (Tushman & Anderson, 1997: v).

However, companies' ability to be innovative is limited by the availability of resources such as financing, skills, know-how and especially time (Meyer, 2001). To overcome these barriers to innovate imposed by the boundaries of the firm (Chesbrough, 2003) there is an increasing trend to externally acquire resources required for innovation via corporate acquisitions (Link, 1988). Cisco's president and CEO, John Chambers puts this new driving force of corporate acquisitions quite simply, 'If you don't have the resources to develop a component or product within six months, you must buy what you need or miss the opportunity' (Bower, 2001: 99). Thus corporate acquisitions help to overcome innovation barriers and open the scope for innovativeness.

The *second new driving force* initiating corporate acquisitions is more long-term oriented. In order to achieve long-term innovativeness and thus sustained profitable growth, the innovation capability has to be based on *profound and efficiently deployed technological competencies*.[9] Such a strong technological resource base is in accordance with the agreed source of competitiveness derived from the companies' core competencies (Prahalad & Hamel, 1990; Sanchez & Heene, 1997; Teece, Pisano, & Shuen, 1997). However, in-house competence building is extremely resource consuming and cannot always entirely cover the need for capabilities[10] required for long-term innovativeness at the right time. Thus companies are increasingly forced to conduct *corporate acquisitions as a substitution to internal competence building R&D* (Bower, 2001; Kwak, 2002).

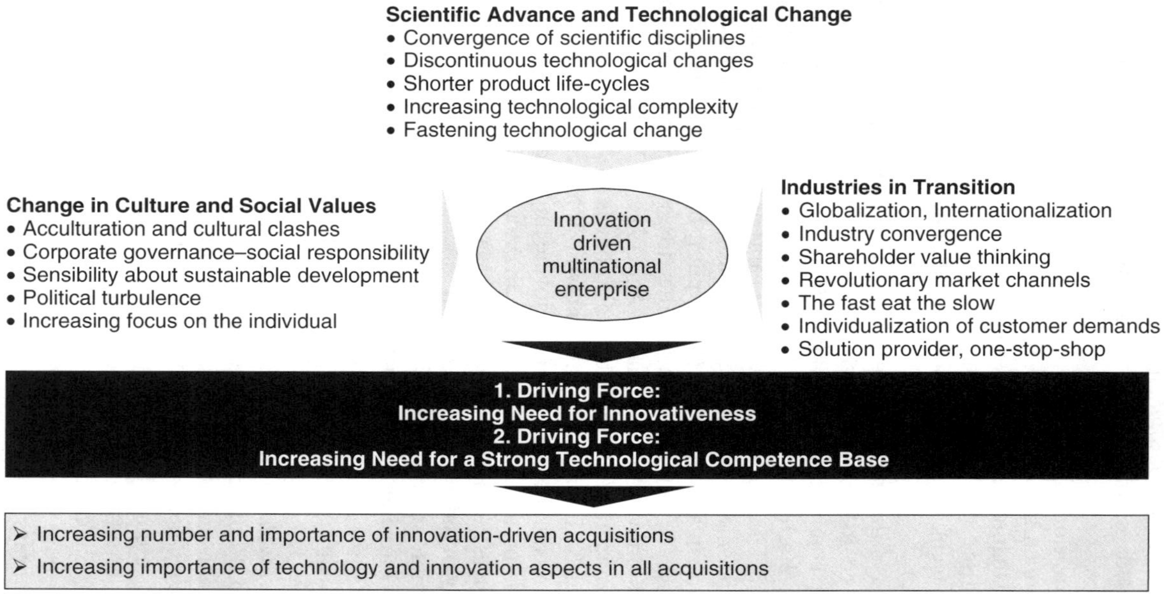

Scientific Advance and Technological Change
- Convergence of scientific disciplines
- Discontinuous technological changes
- Shorter product life-cycles
- Increasing technological complexity
- Fastening technological change

Industries in Transition
- Globalization, Internationalization
- Industry convergence
- Shareholder value thinking
- Revolutionary market channels
- The fast eat the slow
- Individualization of customer demands
- Solution provider, one-stop-shop

Change in Culture and Social Values
- Acculturation and cultural clashes
- Corporate governance–social responsibility
- Sensibility about sustainable development
- Political turbulence
- Increasing focus on the individual

Innovation driven multinational enterprise

**1. Driving Force:
Increasing Need for Innovativeness
2. Driving Force:
Increasing Need for a Strong Technological Competence Base**

⋏ Increasing number and importance of innovation-driven acquisitions
⋏ Increasing importance of technology and innovation aspects in all acquisitions

Figure 1.2 Changes increase the importance of a technology-based perspective on acquisitions

Besides the just described fact that the key drivers in innovation-driven industries result in an increase of innovation-driven acquisitions, the industry dynamics has also an overall influence on all corporate acquisitions taking place within it. The ability to be innovative based on a strong and efficiently deployed technology base are generally crucial for overall long-term success of companies in innovation-driven industries, moreover *this technology-based perspective* on competitiveness has to be *integrated* in all strategic decisions and thus into *general strategic acquisition* and *integration management*. Clemente and Greenspan (1998: 22) also emphasize this new perspective on corporate acquisitions: 'Beyond the obvious attention paid to cost reduction, the primary focus for strategic deals today must be on melding complementary, non-financial assets with an eye toward growth and then extending their benefits over the long term through integration.' This implies that, even in acquisitions aiming for market extension, managers have to increasingly address the acquisitions' impact on the technology base and the companies' future innovativeness.

It can be summarized that technology-based value creation, understood as the short- and especially long-term value creation derived from increased innovativeness and an efficiently deployed resource base, is the determining factor for sustained competitiveness in innovation-driven industries. As a consequence technology-based value creation has gained increasing importance in all strategic acquisitions and has become a main motive and driver for this increasingly important strategic path. Now the question arises whether acquisitions in innovation-driven industries have been successful in achieving technology-based value creation and if the applied understanding of and concepts for acquisition and integration management are appropriate to master the challenges associated with innovation-driven acquisitions.

1.2 Acquisition Performance from a Technology-based Perspective

An analysis of the level of managers' mastery of achieving technology-based value creation in corporate acquisitions soon reveals alarming results. In addition to the well-known fact that acquisitions often fail from a capital market and economic perspective[11] – as is described in several studies[12] – it becomes obvious that they also fail from a technology-based perspective (see overview in Figure 1.3).

For example, studies conducted by Chakrabarti *et al.* (1987; 1994) on managerial perceptions of the success of mergers and acquisitions as a means of acquiring new technologies show that the *professed motives were not achieved*.

Also Chaudhuri and Tabrizi (1999) show that acquisitions with the objective of obtaining capabilities and thus the key to a company's long-term success often arrive at disappointing results. The authors find the reason in managers' shortsighted view of corporate acquisitions. Also

Findings	Authors & Year	Study
Innovation-driven acquisitions aiming for the internalization of competencies **fail to create**	Chakrabarti and Sounder (1987)	31 acquisitions
	Chaudhuri and Tabrizi (1999)	24 high-tech companies, 53 acquisitions
Innovation-driven acquisitions are associated with a particularly **high risk level**	Kitching (1967)	Interviews with top executives of 22 companies on acquisitions in 181 companies between 1960 and 1965. Additionally data from 69 acquisitions
Acquisitions have a **negative effect on R&D input and output**	Hitt et al. (1991a) Hitt et al. (1991b) Hitt et al. (1998b)	191 acquisitions in R&D-intensive industries from 1970 to 1986
Key innovators either leave or lose their innovativeness following an acquisition	Ernst & Vitt (2000)	43 acquisitions of German companies in R&D-intensive industries between 1980 and 1989
Resource deployment in the target's competencies often **result in value destruction**	Capron (1999)	253 horizontal acquisitions of European and US firms in manufacturing industries from 1988 to 1992

Figure 1.3 Overview on acquisition performance from a technology-based perspective

Hitt et al. (1991a; 1991b) note that especially long-term technology-based value creation from acquisitions is rarely achieved.

Kitching (1967) researched the relatedness of acquisition types and their failure rate. Besides the high risks associated with conglomerate acquisitions, where neither the customer nor the technology side are related, the *highest failure rate* was associated with concentric marketing acquisitions aiming for acquiring complementary technologies.

Another influence on the high failure rate associated with the internalization and integration of technologies is the effect of acquisitions on the company's innovativeness. Hitt et al. (1990; 1991a; 1991b; 1998b) pursue the understanding that acquisitions can serve as a substitute for R&D and innovation. They have shown that acquisitions have a negative effect on R&D input, measured in terms of R&D intensity (investment in R&D), and R&D output, measured as patent intensity. This phenomenon is explained by the reduction of managers' commitment to innovation caused by an increasing risk aversion due to the growth, leverage, diversification and size of the company after the acquisition. Furthermore, they describe the negative effect of acquisitions on *the championing culture* of organizational members internally promoting new products and processes.[13]

Another reason for the drop in innovativeness is described by Ernst and Vitt (2000) who show that key innovators after an acquisition either leave or lose their innovativeness.

Last but not least Capron (1999) examines the effect of post-acquisition asset divestiture and resource redeployment on the long-term performance of horizontal acquisitions and shows that especially the changes in the resources of the acquired company result in value destruction rather than creation.

These studies prove that despite the increasing need to innovate based on an efficiently deployed resource base, currently applied acquisition and integration management has so far not succeeded in achieving technology-based value creation.

1.3 Call from Practice – Mastering the Innovation Dilemma

This failure of companies in achieving increased innovativeness and an efficient resource deployment after acquisitions can be referred to as the *acquirer's innovation dilemma*. This dilemma is rooted in managers' difficulties in mastering the specific challenges associated with innovation-driven acquisitions such as:

1. Balancing short-and long-term perspectives: Whereas value creation from market or cost reduction-oriented acquisitions occurs mainly shortly after the acquisition, innovation-driven acquisitions offer long- as well as short-term value-creation opportunities. The challenge lies in appropriately balancing these opportunities and thus in prevailing and managing a short-term and long-term perspective at the same time.

2. Mastering internal growth through external growth: The external growth of a company through an innovation-driven acquisition does not *a priori*, or only in a limited way, lead to value creation. Only the efficient deployment of the resource bases, and thus the development of a new joint competence pool and the creation of a new innovation capability, leads to gains from the acquisition. Thus the main challenge lies in fostering internal value creation through the opportunity of the external growth instead of indulging the acquisition as is.

3. Preserving and using knowledge at the same time: The value competencies and the innovativeness of a target company are dependent on the people having the knowledge and on their context such as working environment, strategic intent, and so on. Thus in order to foster the innovativeness of the target company it should be preserved. On the other hand, technological synergies require the integration of the target's technological resource running the risk of destroying their context and thus their value and the target's innovativeness. The main challenge is to preserve the knowledge and on the other hand to use it to create future value at the same time.

4. Leveraging innovativeness and resource deployment: Technology-based value creation derives from innovation opportunities and the efficient deployment of the resource bases of both companies. The challenge is to understand the dependence of innovativeness on resource deployment, and also the other way round, and to master it in the course of the integration phases.

5. Balancing business as usual and synergy realization: Innovation-driven acquisitions comprise the sourcing of various technology-based synergies as well as existing innovation opportunities. The challenge is to balance the workload and skills to serve the existing customer needs of the individual companies and to successfully realize technology-based synergies.

The reason why these challenges associated with innovation-driven acquisitions are not met, and thus increased innovativeness and an efficient resource deployment are not achieved, lies in two fundamental knowledge gaps in theory and practice. On the one hand, there is a lack of understanding of the mechanisms underlying technology-based value creation and the associated contingencies in acquisitions. On the other hand, the concepts of strategic acquisition and integration management do not sufficiently address the main aspects related to technology-based value creation.

The *lack of understanding* is mainly a result of the previous research conducted within the field of acquisitions. Researchers tend to focus their work on only one specific aspect in corporate acquisitions and relate it to value creation, and do not, however, adequately relate it to other influential factors (Larsson & Finkelstein, 1999).[14] Furthermore, they apply a very short-sighted perspective on acquisitions by using event study methodology[15] rather than investigating the impact factors of the long-term success of corporate acquisitions (Hagedoorn & Duysters, 2002). Additionally, they often do not distinguish their findings according to the acquisition type (Bower, 2001). Thus few implications can be drawn specifically for technology-based value creation. Similarly practitioners in the field of mergers and acquisitions hardly learn from their acquisitions[16] and thus have little knowledge of the aspects relevant for long-term technology-based value creation.

Technology-based value creation is also infrequently considered within concepts, such as processes, structures and tools, of *strategic acquisition* and *integration management*. State of the art in theory lacks concepts which address technology-based value creation and which can be integrated in acquisition and integration management. Accordingly, in practice the people driving the acquisition and integration process are often financial and legal experts or external consultants who prefer to focus on the transaction rather than on potential long-term value creation. Furthermore, the acquisition process is often associated with an escalating momentum, a high level of confidentiality requirements and a strong activity segmentation (Jemison & Sitkin, 1986a) hindering a holistic and integrative approach and the consideration and integration of technology-related aspects. The only concept known is the technology, or technical, due diligence[17] which takes place during the transaction phase[18] of the acquisition process. This, if conducted at all,[19] is mainly designed to validate the condition and maturity of the technologies of the target company or to identify their strengths and weaknesses. Thus technology due diligence is taking a validating role analysing whether the technological competencies were strong in the past. This retrospective approach implies that the future success of the company and thus its associated value creation potential are determined by past and current technological capabilities. Whereas this extrapolating approach might be appropriate to other areas of the company, such as financial resources, it is hardly applicable to technological competencies. This implies that the state-of-the-art technology due diligence is only partly applicable to ensure technology-based value creation in corporate acquisitions. Thus theory and practice lack applicable concepts of the strategic acquisition and integration management which sufficiently address the aspects relevant for technology-based value creation.

It can be summarized that despite the fact that technology-based value creation has become an important issue, and even a driver in corporate acquisitions in innovation-driven industries, it has so far hardly been mastered. Not only do innovation-driven acquisitions fail to deliver value but also general acquisitions have been shown to destroy the potential for technology-based value creation instead of keeping or eventually profiting from it. The main reason for this innovation dilemma lies in managers' lack of mastering the challenges associated with innovation-driven acquisitions. This pitfall in mastering the challenges is rooted in a lack of understanding of the main aspects relevant to technology-based value creation in corporate acquisitions and in the lack of applicable concepts of strategic acquisition and integration management facilitating technology-based value creation. These two gaps, which are persistent in both theory and practice, have so far not found a satisfactory solution.

1.4 Objectives, Research Approach and Content of the Book

The objective of this book is two-fold. On the one hand it aims to explain what technology-based value creation after acquisitions depends on and on the other hand to provide the practitioner with applicable concepts of strategic acquisition and integration management to improve the acquirer's level of mastering the innovation dilemma. Thus the argumentations and discussions are guided by the following two questions concerning strategic acquisitions in innovation-driven industries:[20]

Question 1: Which aspects in strategic acquisitions in innovation-driven industries determine the successful realization of technology-based value creation?

Question 2: How can these aspects be successfully incorporated in the processes, structures and methods of the strategic acquisition and integration management in corporate acquisitions in innovation-driven industries?

In order to appropriately answer these research questions, an applied research approach was taken. In accordance with applied science,[21] this research addresses problems in practice and is clearly aiming to provide solutions to corporate reality with the focus on applicability and usefulness. Ulrich (2001) and others (Schutz, 1973; Lee, 1991) have emphasized that in social or applied sciences versus natural science the hypothesis testing approach is not appropriate, especially as 'people, and the physical and social artifacts that they create, are fundamentally different from the physical reality examined in natural science' (Lee, 1991: 347). Thus, in order to achieve the goal of building a new reality (Ulrich, 2001) and thus answering the first and second research question, an applied and exploratory research approach – the case study research (Yin, 1994) – is favoured.

The decision to apply the *case study approach* for the specific research purpose and topic is justified by the argumentation of various researchers. According to Eisenhardt (1989a: 548), 'building theory from case study[22] research is most appropriate in the early stages of research on a topic or to provide freshness in perspective to one already researched'. This is validated by Kubicek (1975: 61) who confirms that cases are best for the very early stages in research of an organizational problem. Additionally, Yin (1994) describes the case study approach as applying if the researcher has little control over the phenomenon studies and if this is a rather contemporary event. The arguments of Eisenhardt (1989a), Kubicek (1975) and Yin (1994) fully apply to the research topic addressed in this book. Larsson (1993) adds that specifically mergers and acquisitions as highly complex events can be best researched by using the case study methodology.

This case study research approach will be applied to answer the two research questions in the following manner (see also Figure 1.4). Following the recommendations of Ulrich (2001: 213) the research approach is initiated by a problem in practice. Thus in step 1 of this research the call from practice is identified and the associated research topic and questions are determined (see above). In step 2 the background information in the fields of technology and innovation management and of acquisition and integration management are provided. Thus the important terms used within the books are defined and the basic elements of the management of technologies and innovation, and acquisitions and integrations are introduced. In step 3 the case studies will be examined by applying the multi-case study approach. The research model underlying the case studies will be derived, the cases will be selected and their chronological content will be described and analysed, whereas every case serves a specific purpose within the overall scope of inquiry (Yin, 1994: 53). Within step 4 the cross-case analysis is performed and the researcher can derive a generic model of understanding which explains the aspects determining the successful realization of technology-based value creation after corporate acquisitions. The findings from the case studies are complemented by existing literature in this field.

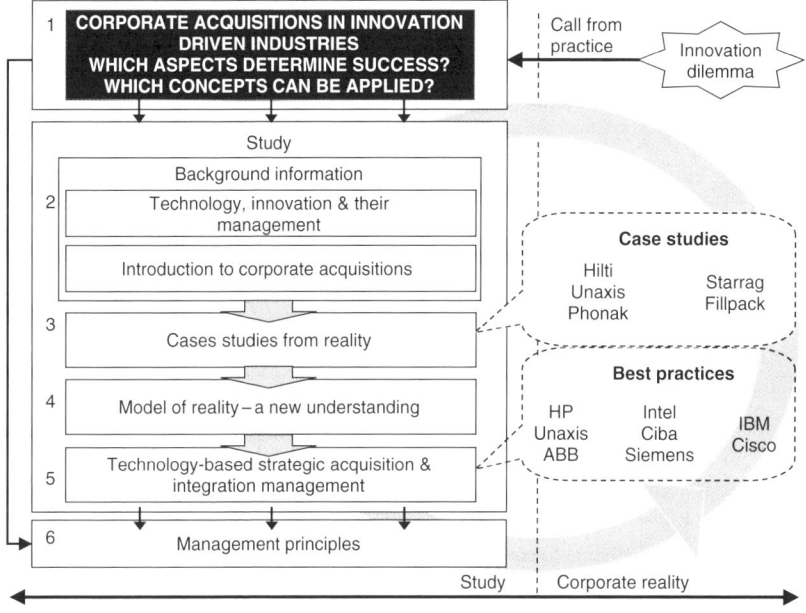

Figure 1.4 Research design of this book

Thus the model helps to answer the first research question posed previously. In step 5 the model on the aspects relevant to achieve technology-based value creation is used to develop a new concept and solution for practice, thus responding to the second research question. This solution will be technology-based strategic acquisition and integration management, which will be complementary, modular and adding to existing concepts in practice. In addition, the solution will be complemented by best practice cases from industry. This concept is one possible solution derived from the model on reality and does not claim exclusivity, even though the researcher aims for a holistic and integrated concept. Finally in step 6 management principles will be developed which purely address management's concern for management style. They are derived from the model on reality and summarize the major aspects relevant to achieve technology-based value creation. Their only objective is to support management in their decision-making process. By following this research approach this book contributes to theory and practice while responding to the two research questions. Accordingly the book is structured in to seven principal chapters following the line of argumentation derived from the research procedure. Figure 1.5 provides an overview.

1 Introduction	Relevance of the topic	Acquisition performance	Call from practice – mastering the innovation dilemma	Objective, research approach and content		
2 Technology, Innovation and their Management	Terms and definitions	Technology management	Innovation management	Conclusion		
3 Introduction to Corporate Acquisitions	Terms and definitions	Acquisition types	Acquisitions as a means of value creation	Acquisitions as a business - strategic path	Conclusion	
4 Case Studies from Reality	Theoretical model to investigate TBVC	Selection of the case studies	Cases from reality	Conclusion		
5 Model of Reality – A New Understanding	Technology-based value creation	Acquisition type	Initial conditions	Strategy processes	External development	Conclusion
6 Technology Based Str. Acquisition & Integration Mgmt.	Developing the concept	Overview of the concept	The Detailed concept	Conclusion		
7 Management Principles	Management principles for a new understanding	Management principles for a new concept	Management principles for implementation			

Figure 1.5 Outline of this book

2
Technology, Innovation and their Management

The first chapter outlined that technology-based value creation has gained increasing importance in corporate acquisitions and needs to be mastered in order to achieve sustained competitive advantage in innovation-driven industries. Furthermore, evidence was provided that technology-based value creation in corporate acquisitions is currently inadequately mastered and an innovation dilemma prevails. Thus the two research questions, relating to improving the understanding of technology-based value creation in corporate acquisitions and to developing applicable concepts for strategic acquisition and integration management, were posed. In the next steps, background information on technology and innovation management is provided. This chapter aims to introduce the reader to an understanding of technology-based value creation and to management concepts for mastering it. Thus in an initial step, the underlying definitions related to technology-based value creation and the factors it depends on are outlined. Secondly, the fields of technology and innovation management are discussed. These fields provide management concepts to master resource deployment and innovation in companies. The provided overview is required to understand what technology-based value creation in corporate acquisitions depends on. Furthermore, technology and innovation management concepts will be integrated into the technology-based strategic acquisition and integration management concepts.

2.1 Terms and Definitions: Technology-based Value Creation

The focus of this book lies on value creation from target and acquirer's joint innovativeness and efficient resource deployment. This value creation is attributed with the term technology-based value creation, or technological synergies, which will be detailed in the following text.

Technology-based value creation is defined as *the short- and especially long-term value creation derived from resource integration mechanisms resulting in innovations and the efficient resource deployment.*[1]

Before explaining the terms innovation and resource deployment, the individual terms of technology-based value creation will be explained.

According to Tschirky (1998c: 226): *'Technologies enclose specific individual and collective knowledge in explicit and implicit forms for product and process-oriented usage based on natural, social and engineering-scientific knowledge.'* He further distinguishes between product and process technologies. While a *product technology* assures that a specific technological impact comes about, a *process technology* enables and/or optimizes the occurrence of the technological impact. In the case of pharmaceuticals, product technology would be the active substance, while high-throughput screening technology applied in research or novel separation technologies applied in production are the drug process technologies.

Value creation is achieved if the *rate of return of an investment is higher than the cost of capital demanded by the capital market.*[2] Value creation is calculated as the sum of discounted cash flows over a determined period of time. In the context of acquisitions value creation is 'a long-term phenomenon that results from managerial action and interactions between the firms. It embodies the outcome of what many people refer to as synergy' (Haspeslagh & Jemison, 1991: 22).

Thus technology-based value creation is the return from the usages of knowledge. This usage resembles resource integration mechanisms which result in innovations and in an efficient resource deployment.

'A firm's *resources* at a given time could be defined as those *tangible and intangible assets which are tied semi-permanently to the firm*' (Wernerfelt, 1984: 172). Furthermore, the resource based view generally distinguishes between two types of resources: tangible input resources such as people, machinery, financial capital; and knowledge-based resources[3] such as organizing principles, skills and processes that direct organizational actions. These knowledge-based resources are characterized by their tacitness[4] (Polanyi, 1966), context specificity[5] (Nelson & Winter, 1982) and dispersion[6] (Weick and Roberts, 1993). These characteristics are often also referred to as fungibility, 'an attribute of a resource that facilitates its application to different organizational and market settings' (Anand & Singh, 1997: 101). As mentioned above, the combination of bundles of resources with the firm-specific context is then referred to as a competence or capability.[7] A specific subset of competencies are *core competencies* (Prahalad & Hamel, 1990). 'Capabilities are considered core if they *differentiate a company strategically*' (Leonard-Barton, 1992: 111).

Resource integration is referred to five different mechanisms to combine or integrate resources to finally create value: resource leveraging, resource fusion, resource reconfiguration, resource transfer and resource substitution. On the one hand the vertical leveraging, fusion[8] and reconfiguration of resources results in the generation of new products and in their commercialization on the market, and thus cause innovations.[9] For example Galunic and

Rodan (1998: 1194) explain the relation between resource reconfiguration and innovation: 'In terms of its source, we can think of Schumpeterian innovation[10] as the reconceptualization of an existing system in order to use the resources from which it is built in novel and potentially rent-generating ways.' These types of innovations based on existing resources are also known as architectural innovations (Henderson & Clark, 1990). The horizontal resource transfer and substitution on the other hand result in an efficiently deployed resource base (see Figure 2.1).

Thus the integration of resources can be seen as the 'flow of competency-related knowledge between competence areas' (Galunic & Rodan, 1998: 1195). These successful knowledge flows and thus technology-based value creation depend upon two specific aspects:[11]

- The availability of combinative capabilities
- The basic characteristics of the knowledge

Successful resource integration and thus technology-based value creation is dependent on the integration capability, also referred to as organizational capability[12] (Grant, 1996) or combinative capability (Kogut & Zander, 1992),[13] which describe the capacity to synthesize and apply current and acquired knowledge. For example, resource leveraging is facilitated by a common understanding and mental models of the cooperating teams. Furthermore, Iansiti (1998) describes that successful innovation and thus resource integration is dependent on experimenting knowledge, knowledge in the technological domains and production systems.

Additionally, resource integration depends on the characteristics of the resources and their social construction within competencies. Galunic and Rodin (1998), for example, show that technology-based value creation is dependent on the tacitness, context specificity and dispersion of the resources to be combined. Further characteristics referring to fungibility are described by Zander and Kogut (1995) or Grant (1996).

Thus value creation from a resource integration mechanism either results from innovations or from the efficient deployment of resources. Whereas resource deployment does not require further definitions, the term innovation needs some clarification.

The term 'innovation' has become very popular.[14] It can be defined as *a company's first successful commercial application of something new*. Generally various types and attributes of innovation can be distinguished. For example Zahn and Weidler (1995) have introduced three different types of innovations which refer to different innovation outcomes. One type is the traditional technological innovation, which includes product and process innovations. Another type is the business-related innovation, for example the introduction of a completely new business model. Thirdly, they have introduced organizational innovation which refers to organizational processes and structures.

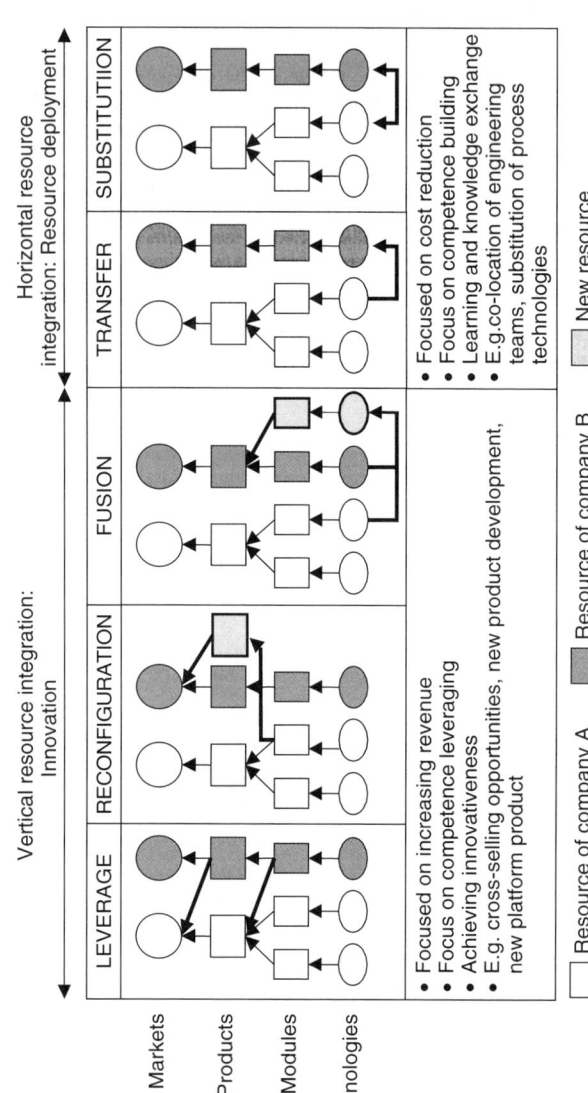

Figure 2.1 Five resource integration mechanisms leading to innovation and efficient resource deployment

Another way to describe innovations is the distinction between incremental, modular, architectural and radical innovations (Abernathy & Clark, 1985; Henderson & Clark, 1990). Whereas within incremental innovations neither the concepts or modules nor their linkages are significantly changed, radical innovations are characterized by new modules as well as new product configurations. The modular innovation has a similar structure to the previous innovations; however, specific modules are overturned. Architectural innovation is characterized by the new configuration of existing modules or technologies. For example, the development of new product platforms is an architectural innovation. These types of innovation also are not confined to technological innovation but apply to organizational and business-related innovation.

Technology-based value creation as part of acquisitions can be incremental, modular or architectural innovations. Radical innovations are rarely an immediate result of corporate acquisitions.

It can be summarized that technology-based value creation based on resource integration mechanisms is the efficient deployment of existing resources and their utilization in terms of innovation. The following section will outline those concepts that can be applied in mastering technology-based value creation, as these need to be integrated into strategic acquisition and integration management. Thus in the later sections the main understanding of technology and innovation management, the schools dealing with technology-based value creation, are introduced.

2.2 Technology Management

Technology-based value creation in corporate acquisitions requires the application of management concepts to efficiently deploy the technologies of firms and to enhance innovativeness. Thus technology-based value creation is facilitated by the concepts of technology management which support the choice and assessment of the appropriate technologies, their transfer and substitution and thus their efficient deployment, and of innovation management which supports the integration, fusion and new configuration of the resources to build new products and services.[15]

Therefore the main concepts of technology and innovation management are introduced here and will be integrated into the technology-based strategic acquisition and integration management concept that is developed in Chapter 6.

2.2.1 Integrated technology management

Generally the purpose of technology management is the deliberate handling of technologies. For decades, several authors have developed various approaches to technology management.[16] This book is based on the Integrated Technology Management theory developed by Professor H. Tschirky (1998c) at the Center for Enterprise Science, Federal Institute of Technology, ETH Zurich.

Integrated technology management is understood as a holistic task of management, which is in line with the normative, strategic and operational objectives of the enterprise and concerned with the design, direction and development of the technology and innovation potential (Tschirky, 1998b: 226). The *technology potential, as a socio-technical subsystem of an enterprise, comprises the available product and process technologies, their personal, informal and material carriers, and the organizational structures and processes required for technological application* (Tschirky, 1998b: 246). The *innovation potential comprises the available innovation competence of individuals and groups at all levels of the organization, which enable organizational inventions regarding the social and technical system, inventions related to the market impact of product and process technologies, and business inventions* (Tschirky, 1998b: 264). This integrated technology management is shown in Figure 2.2. The concept of 'Integrated Technology Management' is motivated by Bleicher's (1996: 223) concept of Integrated Management, adapted for the technological dimension of companies.[17] The vision of the concept of Integrated Technology Management is 'bringing technology into management'. 'Its basis is the postulate that "technology issues" will no longer be solely of concern in the context of direct technology-related managerial functions such as R&D and production management but will be of prime concern for general management at all levels' (Tschirky, 2000: 417). Tschirky details several tasks of technology management which are based on some basic definitions, elements and understandings, outlined in the following.

2.2.2 Basic elements of the integrated technology management

The main and guiding element of *technology management* is technology strategy. Tschirky (1998a: 293) states that the purpose of technology strategies is twofold: on the one hand, technology strategies draw up a solid foundation for decision making in order to enable the selection of technologies and strategic technology fields that are suitable for the creation and maintenance of an enterprise's competitive position.[18] This aspect is equivalent to the notion of a technology strategic goal or objective. On the other hand, technology strategies have to illustrate the appropriate paths leading to the mastery and deployment of the selected technologies.

The main content of the technology *strategic goals* is the determination of which technologies will be deployed and used within the company. In 'technology management language' technology strategic goals describe which technologies of a strategic technology platform will be integrated in the strategic business fields of the company. A *strategic technology platform* reflects an aggregation and bundling of product and process technologies, required application know-how, and underlying theories from a strategic point of view, which – with regard to the complexity inherent in many technologies – increases the manageability of an enterprise's technological knowledge

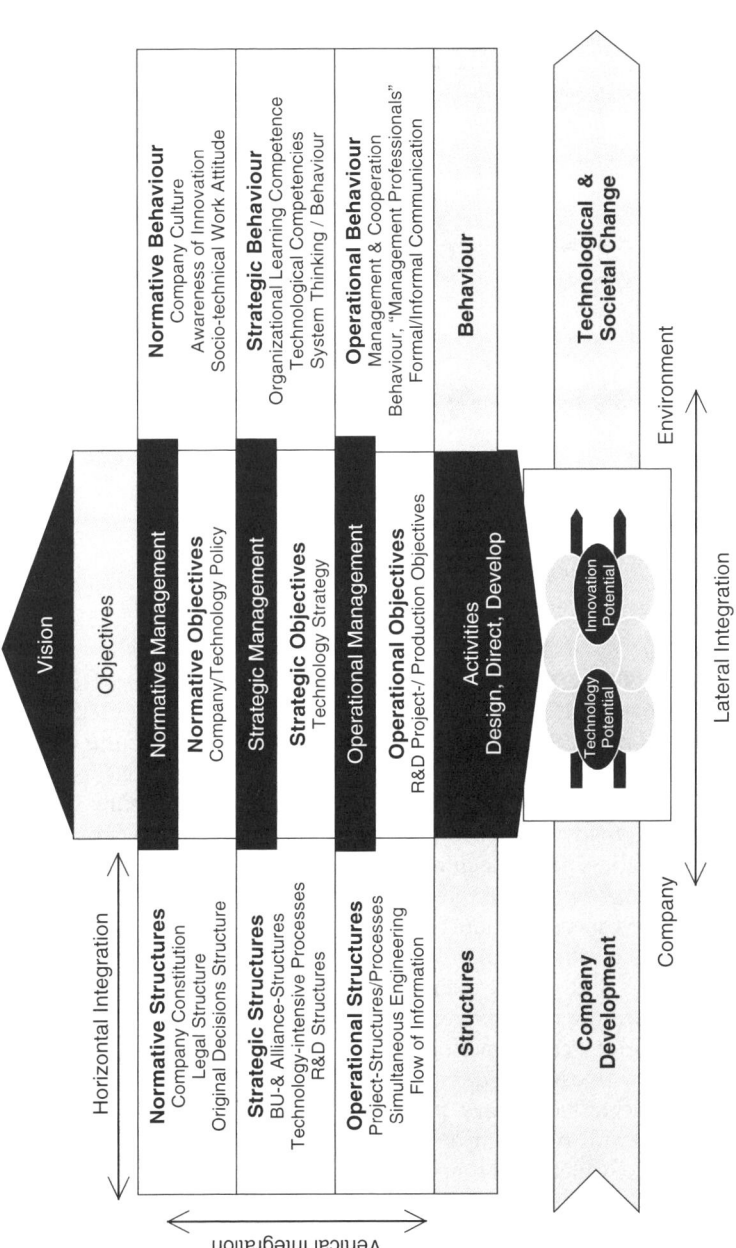

Figure 2.2 'Integrated Technology Management'
Source: Tschirky (1998b: 270).

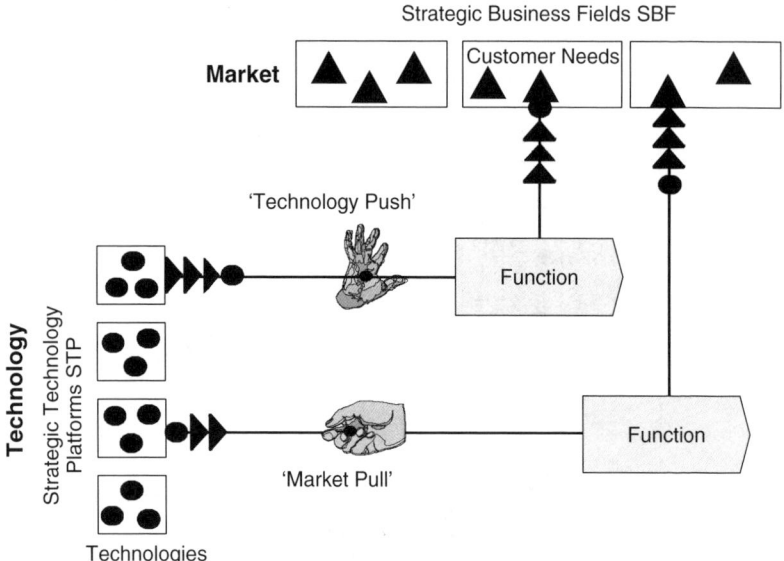

Figure 2.3 Functional match between technologies and needs
Source: Tschirky, Jung and Savioz (2003: 71).

(Tschirky *et al.*, 2003). Strategic technology platforms often comprise a core technology representing the platform. It fulfils a main function used to satisfy a customer's needs. Strategic technology platforms are the counterpart to *strategic business fields* which assemble knowledge on specific markets, their relevant customers, products and services. Customer needs are defined within strategic business fields. The relation between strategic technology platforms and strategic business fields is achieved through functions (Figure 2.3).

Generally it can be presumed that a customer does not express his or her needs in terms of a specific product (this is only the case if the customer aims for a certain brand rather than a certain value proposition) but in terms of specific functions and related requirements. For example, a customer who wants to buy a drilling machine requests a function such as that the drill be able to drill a hole at a certain performance. These customer needs, translated into a function with associated requirements, are fulfilled by the technologies of the strategic technology platform. Thus technology management aims to develop and maintain distinctive core technologies within these strategic technology platforms that can be leveraged and used across multiple strategic business fields. This functional match between customer needs within the business fields and technologies within the technology platforms are driven by two distinctive forces. One force originates from the market side, such as from increasing customer needs, fierce competitions or a new regulation. This force is also called *market pull*.[19] The other force initiating a functional match is called *technology push*. This represents the urge to integrate new technologies into products and services for the customer. Whereas responding to the market pull with new products is a reactive way to satisfy customers, a technology push approach can generate new market needs.

One practical tool, which outlines the relations between strategic business fields and technology platforms matched by functions is 'Innovation Architecture' (see Figure 2.4). It will be used throughout the book as a representation of the knowledge within a company.

Once the strategic goals have been defined the appropriate *technology strategic path* has to be determined. Generally the decision upon the strategic

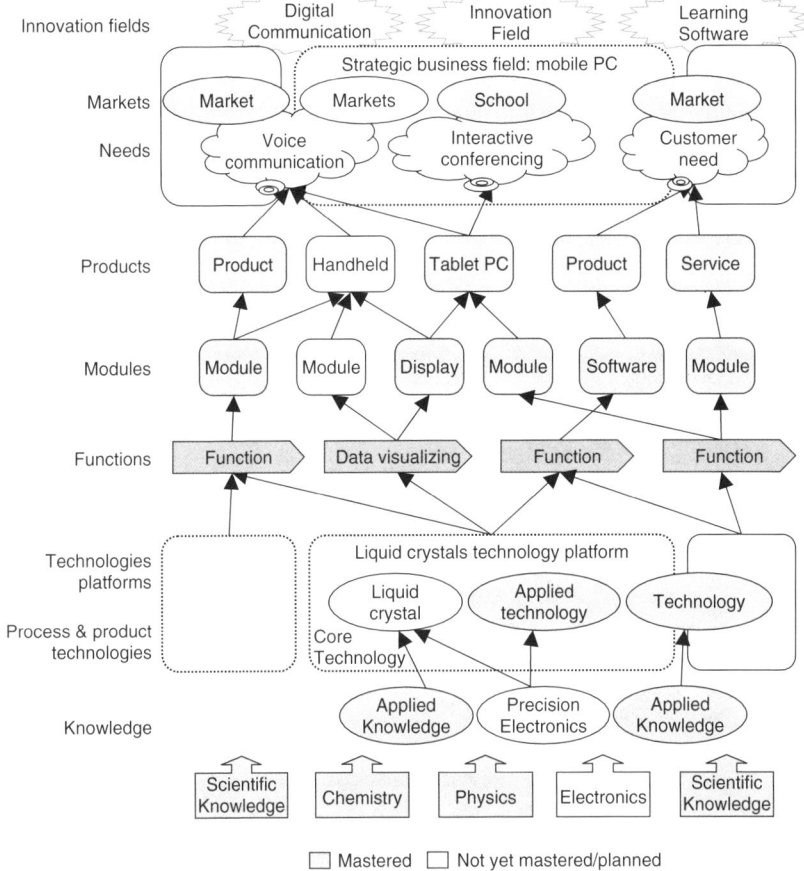

Figure 2.4 Innovation architecture

Source: Sauber (2003).

path is concerned with *three main strategic decisions which address the deployment of technologies*:

1. Which technologies are required ('Which Technologies')?
2. Shall the technologies be made internally or externally acquired ('Make-or-Buy')?
3. Shall the technologies be kept or sold ('Keep-or-Sell')?

The first decision of this trio is concerned with the demand for specific technologies. Based on technology strategy, which determines the functional requirements and core technology fields, the strategic technology decision on 'which technology' investigates which familiar, unfamiliar or completely new technologies should be deployed. The second decision, make-or-buy, is concerned with the question as to whether the required technologies are to be made available through acquisition, collaboration with other companies or through in-house development. Thus this make-or-buy decision will be an integral part of the decision to acquire a company. The third decision is concerned with the already available technologies within the companies. It determines whether the technologies will be confined to internal use only or will be exploited on the market for knowledge and technologies. These three decisions are tightly interdependent and together represent the 'trio of strategic technology decisions' (Figure 2.5).

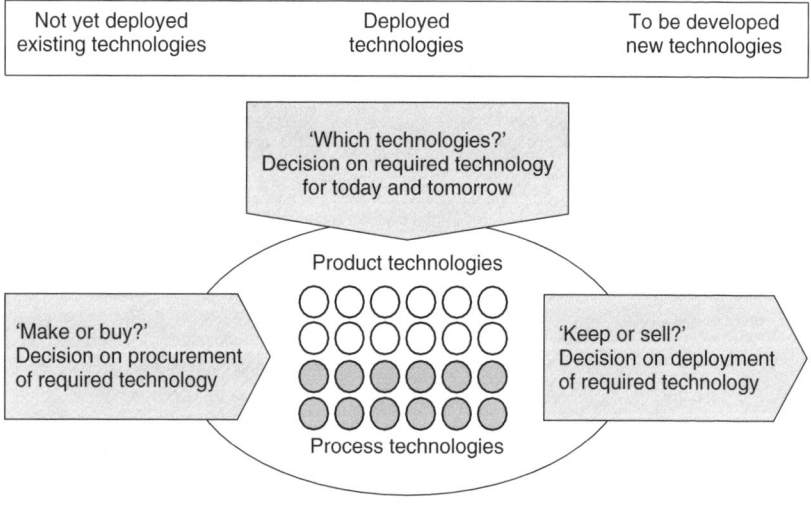

Figure 2.5 Trilogy of strategic technology decisions
Source: Tschirky and Koruna (1998), after Brodbeck *et al.* (1995: 108).

2.2.3 Main tasks of the integrated technology management

In order to support the formulation and implementation of technology strategic goals and paths, the technology management is segmented into different tasks. Tschirky (1998b) provides an overview of the main tasks of technology management on the normative, strategic and operational level. The most important tasks which are also related to and can be used within corporate acquisitions are outlined in the following. Figure 2.6 shows the different tasks which support technology strategy formulation and implementation processes. In the following, strategic technology planning, technology assessment and controlling, technology marketing and intelligence are briefly introduced.

Strategic *technology planning* has attained much attention in the recent past; however, no common frame and definition on that subject has evolved so far. Whereas some authors subsume the whole technology strategy formulation and implementation tasks and processes under the term strategic technology planning,[20] others include only technology intelligence, assessment and planning,[21] and again others see technology planning as a discontinuous process to generate a plan of the intended technology development and use.[22] Within this book technology planning is viewed in the narrow sense especially as the other topics such as technology intelligence and assessment are discussed separately. Thus strategic technology planning is defined according to Bucher (2003), who refers back to Mintzberg as

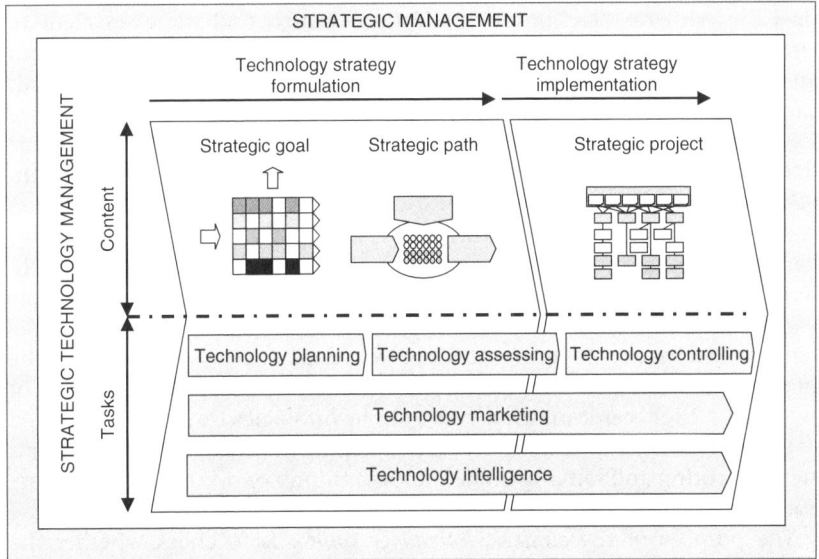

Figure 2.6 Technology strategy formulation and implementation and associated tasks

a formalized procedure to produce an articulated result, in the form of an integrated system of strategic technological decisions (Bucher, 2003: 69). The strategic technology planning process is based on the strategy formulation process of the strategic planning school, as a prescriptive and formalized process.[23] This generic strategic planning process consists of steps such as (1) the strategy identification, (2) environmental analysis, (3) resource analysis, (4) gap analysis, (5) strategic alternatives, (6) strategy evaluation, (7) strategic choice. Tschirky (1998b: 295) has proposed to integrate strategic technology planning into this generic business strategic planning process. He aims to ensure collaboration and strategic consistency between business fields and technology platforms.

Technology assessment is a highly challenging topic in technology management and acquisition and in integration management especially as the value of partly tacit and dispersed knowledge, such as applies to technologies, has to become comparable or even directly expressed in facts and figures. Technology assessment is used to conduct a rough company evaluation within the technology intelligence, to decide upon strategic technology options, to assess the market value of a technology to be sold or purchased via technology marketing and to conduct the budgeting of technology projects.

Technology assessment takes place on various levels, depending on the risk associated with the technology and thus also with its maturity.[24] It can be distinguished between the assessment of technology projects with a determined application context, the assessment of technologies within a firm but with no determined application context and the assessment of emerging technologies, whereas neither the role within the firm nor within an application are known. Furthermore, technology assessment methods can be categorized according to quantitative and qualitative methods[25] (see Figure 2.7). Highly certain technology projects, especially following the dominant design,[26] can be assessed via qualitative and quantitative methods. For example, the feasibility of a certain technology in an application context can be determined or the project can be assessed on a quantitative basis by using the discounted cash flow methodology.[27] If the focus of a technology not only rests on one particular application but is seen as a valuable contribution to the whole company, then methods such as the core competence model become more viable. The core competence method analyses the fit of the technology to the core competencies and thus the strategic focus of the firm.[28] Highly uncertain or emerging technologies cannot be evaluated as projects and in a quantitatively way. In this case more qualitative methods such as scoring and ranking model, rules of thumb or portfolio methods are applied.

The purpose of the *strategic technology control* is to check whether the technology strategy is implemented as planned and if the results in terms of new technologies, a higher innovativeness or the successful acquisition of a

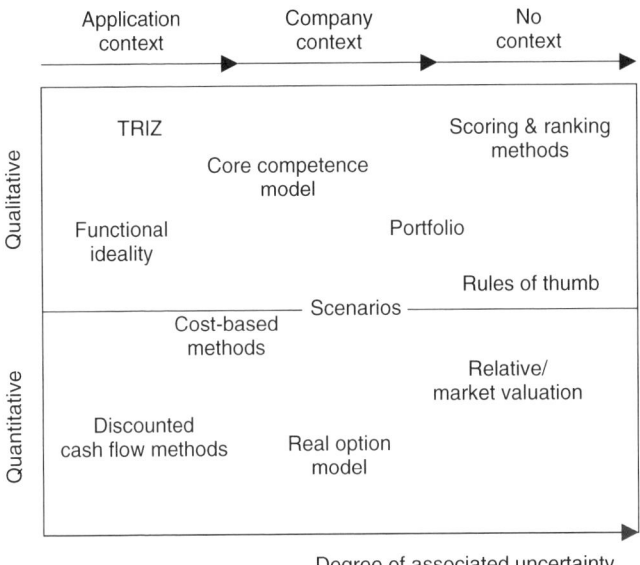

Figure 2.7 Technology assessment methods depending on degree of uncertainty

technology are those intended. Several authors have introduced various tools to support strategic technology controlling on all levels, such as strategic R&D projects, milestone analysis, technology calendars, technology portfolios, innovation rate checks, cockpit-charts, and so on.

Technology intelligence includes activities that support decision-making of technological and general management concerns by taking advantage of a well-timed preparation of relevant information on technological facts and trends (opportunities and threats) of the organization's environment by means of collection, analysis and dissemination (Savioz, 2002: 36). The purpose of technology intelligence is manifold,[29] however always somewhat related to the anticipation and understanding of science and technology-related trends or discontinuities and to the awareness of potential technological threats.

Once the decision is made about which technologies are required for future business within the strategy formulation process (see Figure 2.6), the decision is made as to which technologies will be developed internally or acquired and which will be kept and exploited internally or sold (see trilogy of technology decisions above). Thus these strategic decisions on the path to achieving the technology strategic goal are the driving forces of *technology marketing*. Technology marketing is concerned with technology acquisition and external exploitation or deployment. Thus such topics as IP management and licensing, technology spin-outs, joint ventures, and also technology-intensive acquisitions are subject to technology marketing.

It can be summarized that strategic technology management is mainly concerned with the formulation and implementation of technology strategy. In doing so, different tasks such as technology planning, assessment and controlling as well as the continuous tasks of technology intelligence and marketing support the strategic processes. These management concepts to master an efficient resource deployment will be integrated within the strategic acquisition and integration management to foster the acquirer's mastery of the innovation dilemma.

2.3 Innovation Management

The previous section on Technology Management has introduced concepts to support technology-based value creation from resource deployment. Innovation as a subsequent step is the utilization of the diligently selected and assessed and efficiently deployed resource base to create value through innovations. Thus the main notions of, and concepts to manage, innovations have to be outlined and incorporated into the technology-based strategic acquisition and integration management.

Innovation management also comprises innovation strategy formulation as well as implementation. However, in current innovation management research the focus lies on strategy implementation, in particular on the innovation process which guides innovation strategy implementation. As Hauschildt (1993a: 25) describes, innovation management deals with the organization or coordination of the innovation process. Literature provides numerous, varying views on the innovation process, how it is organized and structured in phases and milestones, stages and gates.[30] Common to all of them is that at the front-end there is something like an idea, and at the end a kind of realization and commercialization of the idea.

Thus technology-based value creation in corporate acquisitions has to be realized as innovation projects, following the innovation process. The acquisition integration process from a technology perspective can be seen as the accumulation of various innovation projects which need to be mastered. Unfortunately there is hardly any literature on innovation processes which are based on completely different knowledge bases, as is the case in acquisitions. Furthermore, cultural differences and other acquisition-specific aspects in innovation projects have hardly been addressed. Thus, so far, few concepts besides innovation processes can be adapted from innovation management and integrated into technology-based strategic acquisition management.

This review of technology and innovation management has provided an overview of concepts supportive to innovations and efficient resource deployments. These will be adapted and integrated into strategic technology-based acquisition and integration management.

2.4 Conclusion

This chapter has introduced the basic terms and definitions related to technology-based value creation and has introduced management concepts to master efficient resource deployment via technology management and to achieve innovativeness through innovation management. This background information will be applied throughout the subsequent chapters of this book. Thus definitions of technology-based value creation mechanisms will be used in describing the selected case studies and in explaining which key factors have impact on the likelihood of their occurrence. The elements and management concepts of technology and innovation management will be integrated into technology-based strategic acquisition and integration management and thus facilitate the appropriate mastering of the innovation dilemma.

3
Introduction to Corporate Acquisitions

After having reviewed the underlying concepts required for mastering an efficient resource deployment and innovativeness, the fundamentals on corporate acquisitions are addressed. First of all, the basic terms and definitions related to corporate acquisitions are outlined. Subsequently, the different types of acquisitions and their dual and interrelated purpose as a means of creating value and of achieving a strategic objective are introduced. The objective of this chapter is to provide the reader with background information on acquisitions which are useful within the subsequent discussions on acquisition and integration management.

3.1 Terms and Definitions: Corporate Acquisitions

The terms 'corporate acquisitions' or 'mergers and acquisitions', often referred to as 'M&A', have become buzz-words in financial institutions and strategic management groups rather than well-defined terms. Copeland and Weston, for example, state that 'the traditional subject of M&As has been expanded to include takeovers and related issues of corporate restructuring, corporate control, and changes in the ownership structure of firms' (Copeland & Weston, 1988: 676). Within this book, following a wide stream of researchers on corporate acquisitions,[1] mainly three methods for combining two companies are considered as corporate acquisitions: asset deals, share deals and mergers through fusion (see Figure 3.1). This implies that within this book the terms 'merger' and 'acquisition' are used interchangeably, a practice common for research which does not specifically focus on mega-mergers.

Generally the transaction can be structured as a share or as an asset deal.[2] Whereas within a share deal the ownership and thus the corporate control over the target company is exchanged,[3] within the asset deal the individual assets are sold to the acquirer.[4] The share deal is only applicable when the target is a company, whereas the asset deal can comprise the acquisition of sub-units of a company. Share deals are relatively unproblematic; however

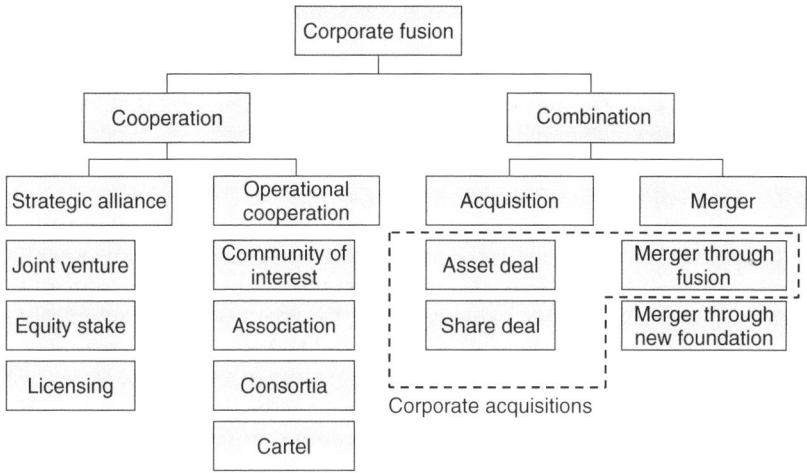

Figure 3.1 Corporate acquisitions
Source: adapted from Berens & Brauner (1999: 52).

they are associated with disadvantages in comparison with the asset deal, especially in tax and financing issues. The asset deal, on the contrary, has to be favoured from the tax perspective; however its process is highly complex.[5]

3.2 Acquisition Types

Acquisitions can be categorized according to the relation between the acquired business and the existing business of the company. This categorization can be done on an industry and on a business level.

The relation on the *industry level* is defined by horizontal, vertical, concentric or conglomerate acquisitions. At horizontal acquisitions acquirer and target are in the same industry. They serve the same customers and either differ or are similar in their product offering. Thus a horizontal acquisition can be used to deepen or extend the product range. At vertical acquisitions the target company is either a supplier or a customer of the acquirer. At concentric acquisitions[6] the target company has either a similar customer base but different product and technologies or a similar technology base but different products and customers. This type of acquisitions supports the potential to leverage market access or competencies. At conglomerate acquisitions the target company is in a different value chain, serving different customers and having different suppliers.

The relation on the *business level* describes how the target company is related to the current set of businesses. Haspeslagh and Jemison (1991), for example, distinguish between domain strengthening, domain extension and domain exploration. Ansoff (1965) has categorized similarly into acquisitions which extend the market side of the company or the technological side or even both domains of the company.

All these types of acquisitions aim for a two-fold purpose: to allow value creation and to contribute to a certain strategic path.

3.3 Acquisitions as a Means of Value Creation

Generally the main driving force in acquisitions is the search for potential competitive advantage derived from the combination of two companies. This potential shall enable value creation, the main objective of entrepreneurial actions, and thus increase the shareholder value.[7] There are different ways that acquisitions can create value, such as by increasing operating cash flow, reducing the cost of capital or by improving the financial leverage. The focus of this book lies on strategic acquisitions primarily aiming for an improvement of the operating cash flow.

3.4 Acquisitions as a Business Strategic Path

As described, value creation derives from an improved competitive advantage of the merged companies versus operating as two individual companies. Thus corporate acquisitions can be seen as a strategic path to achieve an increased competitiveness. Haspeslagh and Jemison (1991), for example, distinguish three different types of an acquisition's contribution to the companies' strategies:

(1) Acquiring a specific capability
(2) Acquiring a platform
(3) Acquiring an existing business position

These three types of contributions vary in the extent to which their business strategy is fulfilled. Whereas the first contribution adds only very little, acquiring a platform represents a clear commitment to an investment strategy that far exceeds the initial purchase price of the target. Acquiring an existing business position is equivalent to fully implementing the pursued strategy.

Other authors[8] have discussed the role of an acquisition to achieve competitive strategies, such as strategies derived from portfolio management, from the resource-based view, from the technology-based approach or from the positioning school of the five forces by Porter. Additionally, Bower (2001) distinguishes various motives without referring to a specific strategy: the overcapacity M&A, the geographic roll-up M&A, the product or market extension M&A, the M&A as R&D, and the industry convergence M&A.

It can be summarized that acquisitions can serve as a strategic path for several strategic goals. Technology-based value creation becomes most important in acquisitions derived from a technology strategy, as the strategic goals comprise the realization of technological synergies. Nonetheless, technology-based value creation can be achieved in different types of acquisitions, driven by the various motives.

3.5 Conclusion

Within this introduction to corporate acquisitions the main definition of acquisitions as share or asset deals or mergers was provided. Furthermore, different acquisition types depending on the industry or business relation of the joining companies were introduced. Additionally the dual purpose of acquisitions as means to create value and to achieve a strategic goal was discussed. Innovation-driven acquisitions aim for technology-based value creation and the strategic goal to internalize competencies and raise innovativeness. These explanations will be used within the subsequent chapters of this book.

4
Case Studies from Reality: Technology-based Value Creation in Real-life Acquisitions

Within the last three chapters it was outlined that an increased innovativeness and the efficient deployment of resources are gaining importance in corporate acquisitions; however, they are not yet mastered. Thus this book has been initiated to better understand the relevant aspects impacting on technology-based value creation in corporate acquisitions (Research Question 1) and to develop applicable concepts which can be integrated into strategic acquisition and integration management to increase its level of mastery (Research Question 2). In order to support this interest, background information on technology and innovation management and on corporate acquisitions was provided in the last two chapters. In this chapter practice is consulted to answer the first research question which was posed in Chapter 1. Thus this chapter aims to identify the aspects relevant to achieve technology-based value creation in corporate acquisitions. This objective is reached by applying the case study methodology according to Yin (1994).

Thus for the first step, a theoretical model, or theory which resembles the investigation, has to be developed. This theoretical model, which is based on existing literature, indicates which aspects need to be analysed within the case studies to answer the first research question. In the second step, the cases which match the frame of investigation are selected and the data collection protocol designed. In the third step, the cases are described in chronological order, analysing the aspects of the developed theory. In a subsequent chapter, the cases are compared, the theoretical model is modified and a new understanding is derived. Finally, implications are drawn for practice by developing the technology-based strategic acquisition and integration management and by deriving management principles (see also Chapters 6 and 7).

Thus this chapter contains four sections. The first describes the theoretical model for investigating technology-based value creation in corporate acquisitions. The second outlines how the cases were selected and how the data collection was designed and the third section comprises the description of the case studies. The fourth section is a brief conclusion to the chapter.

4.1 Theoretical Model to Investigate Technology-based Value Creation

The theoretical model should be a framework of aspects relevant to achieve technology-based value creation in corporate acquisitions derived from the state of the art in literature. However, literature so far hardly addresses technological synergies in corporate acquisitions. Researchers have mostly focused on general value creation and tried to derive the relevant aspects impacting on it rather than analysing the key success factors for innovativeness after acquisitions. Thus the theoretical framework will be derived from research on general value creation in acquisitions and applied to the innovation-specific context. Generally four different research schools have investigated the aspects relevant for general value creation in corporate acquisitions:[1] the capital market school; the strategic school; the organizational and behavioural school; and the process school. The theoretical model can be derived from researchers in these fields (for a detailed investigation of these research streams see Appendix A).

From these research publications the following framework on the different aspects relevant to value creation in corporate acquisitions can be derived (see Figure 4.1). Generally three different groups of influencing

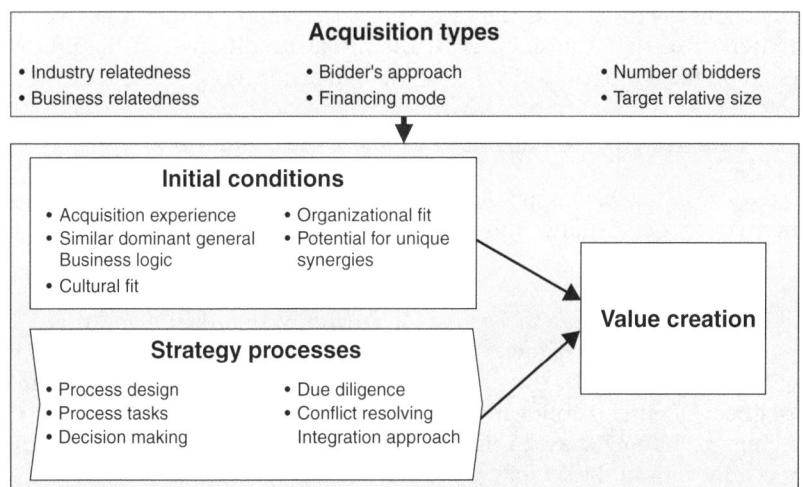

Figure 4.1 Influential aspects on the successful realization of value creation

factors were identified:

(1) The acquisition type
(2) The initial conditions
(3) The process design.

Thus value creation in acquisitions depends on the characteristics of the acquisition, such as the relatedness of the target and the acquirer, their relative size, the bidder's approach and financing mode, as well as the number of bidders. Furthermore, the likelihood for success in acquisitions is determined by the initial characteristics of target and acquirer, including the acquirer's acquisition experience, the fit between the two companies and the potential for unique synergies. The most important impact factors, however, are the design, structure and tasks of the acquisition and integration processes. Thus researchers have found that the way the processes are managed, structured and led highly determines whether an acquisition will create value or not.

This framework helps to understand and especially further investigate the aspects that are important for technology-based value creation in corporate acquisitions. The main pitfall of this framework so far is that the aspects have been linked to neither technology-based value creation nor each other. Thus researchers have so far hardly attempted to link the different acquisition types, and especially motives, to the other impact factors and finally to the pursued value creation. The framework serves well as a generic model; however, key criteria relevant for technology-based value creation cannot yet be derived. However, the framework can be used to discuss the case studies to investigate which of these aspects are particularly relevant for technology-based value creation. Thus the case studies that follow will be analysed for the impact of the acquisition type, the initial conditions and the process design on technology-based value creation in acquisitions.

4.2 Selection of the Case Studies and Data Collection

To retrieve data which apply to the theoretical model and help to build an improved understanding, the cases have to be selected according to the following criteria.

(1) The acquirer has to be part of an innovation-driven industry: As described in the first chapter, the urge for innovativeness and an efficiently deployed resource base is especially crucial for competitiveness and sustained growth in innovation-driven industries.[2] Thus the objective of reaching technology-based value creation more likely will be a crucial issue for acquirers from these types of industries.

(2) The acquisition must be strategic: As mentioned before, there are various transaction types grouped under the term acquisition. In strategic

acquisitions, as opposed to management or leveraged buy-outs or acquisitions which aim for improved financial leverage, the acquirer hopes to achieve a certain business strategic objective and to increase the company's operational cash flow, competitiveness and value. These strategic acquisitions can aim for improved market access, a new business opportunity, the internalization of technological competencies or the increase in economies of scales or scope.[3]

(3) Technology-based value creation has to be a crucial factor for acquisition success: This research is not restricted to acquisitions which primarily aim for technology-based value creation, as is the case in innovation-driven acquisitions. The main reason for widening the scope of this research to all strategic acquisitions where technology-based value creation is an important issue comes from the argument at the beginning of this book, claiming that technology aspects are not only relevant to acquisitions which aim to only internalize technologies but to all corporate acquisitions in innovation-driven industries. Furthermore, in the course of this research it soon became clear that especially in acquisitions primarily aiming for increased market success, technology-based value creation could be a crucial source of gains and success. However, despite these potential gains, technology aspects are mostly neglected in acquisitions aiming for market access, which can have very negative impact on the acquisition performance. For this reason the scope of this research is not confined to innovation-driven acquisitions. It can be argued that in acquisitions in innovation-driven industries which have a shortsighted focus on gaining certain market channels or the like, technology-based value creation is neither an issue nor relevant for the companies' success. Therefore, it is decided that all strategic acquisitions, which see technology-based value creation as one of the three major objectives or sources of value creation, match the theory developed and research applied within this book.

Intentionally there is no specific indication of the relative size of the acquisition target to the acquirer. This is due to the fact that the impact of the acquisition type which includes the relative size of the target is one variable of investigation. However, to limit the complexity of this research in favour of the quality and comprehensiveness of the derived model on reality, the recently occurred mega-mergers, such as the fusion of Vodafone and AirTouch Communications Inc. in 1999, and, more recently, in October 2003 the fusion of General Electric's and Vivendi Universal's entertainment empires, which created a $43 billion entertainment powerhouse with assets ranging from Hollywood's Universal Studios to US television network NBC, are not addressed within this book.

Following these criteria, five case studies were selected to be introduced within this book. Each case is concerned with a specific acquisition. Besides these, the author was in contact with several other companies discussing the relevance of the topic, experiences from past acquisitions and their approach towards strategic acquisition and integration management, and especially technology-based value creation within it. The findings from companies such as Intel Corporation, Siemens Corporation, IBM Corporation, BASF, Cisco Corporation, Hewlett Packard, Ciba Specialty Chemicals, Gurit AG and others, will be included within the further development of the solution concept introduced in Chapter 6.

As this research is highly subjective due to the direct and interactive way of research and due to the scope of interpretation provided for developing the model of understanding, the full and objective truth will not be described. Thus the reader has to be aware that the subsequently derived model of understanding is an interpretation of various factors linked to each other. Despite the natural limitations of this exploratory research approach, a comprehensive and somewhat holistic perspective on technology-based value creation in corporate acquisitions is attempted.

Besides merely explaining the findings derived from the case studies, these will be complemented by other related findings from literature. Thus the soundness of the argument is either increased or challenged and the reader can judge its relevance for him- or herself.

The cases were conducted over a one- to two-year period, at least one or two years after the closure, which allowed not only the observation of the integration progress and related shift in opinions over time but also a return to the people involved in the cases and to retrieve additional data and verify emergent patterns and findings.

The cases were mostly conducted on site via interviews with one to four people at the same time. The interviews mostly took between one-and-a-half and four hours.[4] The data for each case were collected from a minimum of four interviews with people such as the members of the corporate M&A team, chief technology officers, chief executive officers, financial officers, strategic planners, product line managers and engineers. In most cases the author tried to conduct interviews with representatives from both sides, that is from the acquirer and the acquired company. However, due to external influences, location distance and other factors, this aim could not always be achieved. The interviews were designed semi-structurally and followed an open questionnaire. Besides the data from the interviews the author was provided with several additional information such as internal presentations, meeting notes, internal reports, financial calculations, and the like. Especially as the topic of corporate acquisitions is often highly sensitive and only known to the top strategic management, the information gathered is often not public and thus cannot be published. In these cases the issues are either described on a highly abstract level or even circumvented. Information disclosed within this book is approved by the contact persons at each company. Only one company in the packaging industry did not want to disclose its name and the accurate figures, thus the name of the acquirer has been changed to Fillpack and the name of the target became

Aseptofill; additionally the figures were adapted. The names of the other companies are actual.

Each case follows the same structure. It is described according to chronological order. Depending on the focus of the case the different phases are discussed more or less extensively. The case study description is initiated with a brief overview of the acquisition and its key figures. Then the acquirer and its initial situation before the acquisition is described and the development of the acquisition idea and the finding of the potential target are outlined. Next, the target company, its activities and technologies are described. Then the acquisition and the integration processes and the achieved technology-based value creation are described. Each case is concluded with a discussion of the aspects relevant for technology-based value creation. This discussion is led by the framework identified in literature which indicates that the acquisition type, the initial conditions and the strategy processes significantly determine the potential for technology-based value creation. Interestingly in the course of the case study research another aspect, so far hardly mentioned within literature, proved to influence the potential for technology-based value creation. It was observed that the external developments also have to be considered in explaining the factors that impact on technology related value creation. Thus the framework for discussing the cases is expanded to include this aspect.[5] These factors and their influence are discussed individually and also interrelatedly.

4.3 Cases from Reality

Each case study contributes to an improved understanding of technology-based value creation in a specific way. The cases are ordered according to the relative size of the target to the acquirer and also according to the relevance of technology-based value creation within the acquisition. An overview of the specific attributes of the five cases can be seen in Figure 4.2.

4.3.1 Case 1: Hilti–Ammann Lasertechnik acquisition

The Hilti Corporation based in Schaan, Principality of Liechtenstein, acquired the Ammann Lasertechnik Company, Switzerland in February 2001 (for the acquisition timeline see Figure 4.3). The acquisition aimed to acquire and internalize development competencies for the 'positioning business' unit within Hilti. Whereas the acquired company and especially its owner strongly contributed to the development of the next product generation, the internalization and transfer of engineering knowledge and the long-term integration of competencies required special effort, mainly because it was embedded as tacit knowledge within the target's owner.

4.3.1.1 Initial situation

The Hilti Group was founded by Martin Hilti in 1941 as a machine shop in Schaan, Principality of Liechtenstein, and is now a world leader in developing,

		Buying company					Target company					Acquisition characteristics		
Case	Industry	Company/ Unit	Year of figures	Revenue (in CHF m)	Empl.	% R&D invest.	Country	Company/ Unit	Revenue (in CHF m)	Empl.	% R&D invest.	Country	Price (in CHF m)	Date of signing	Acquisition motive
1	Building	Hilti	2001	3138	13780	4.4	LI	Amman Laser-technik	6	10	N/A	CH	N/A	2001	Competencies
2	Packaging	Fillpack	1998	2000	6000	6–7	DE	Aseptofill	50	400	3	DE	25	Jan-1999	Completing product range, competence & market access
3	Hearing aid	Phonak AG	2000	314	1280	6–7	CH	Unitron AG	110	650	3–4	CAN	161	Nov-2000	Initially brand & market, then technology
4	Semi-conductor	Unaxis/IT Division	1999	83	N/A	N/A	CH	Plasma-Therm Inc.	61	180	N/A	USA	240	Dec-1999	Complementary technology, market access & scale
5	Machining	Starrag AG	1998	70	400	10–12	CH	Heckert	90	380	5–7	DE	N/A	Aug-1998	Market access & competencies

Figure 4.2 Overview of case studies

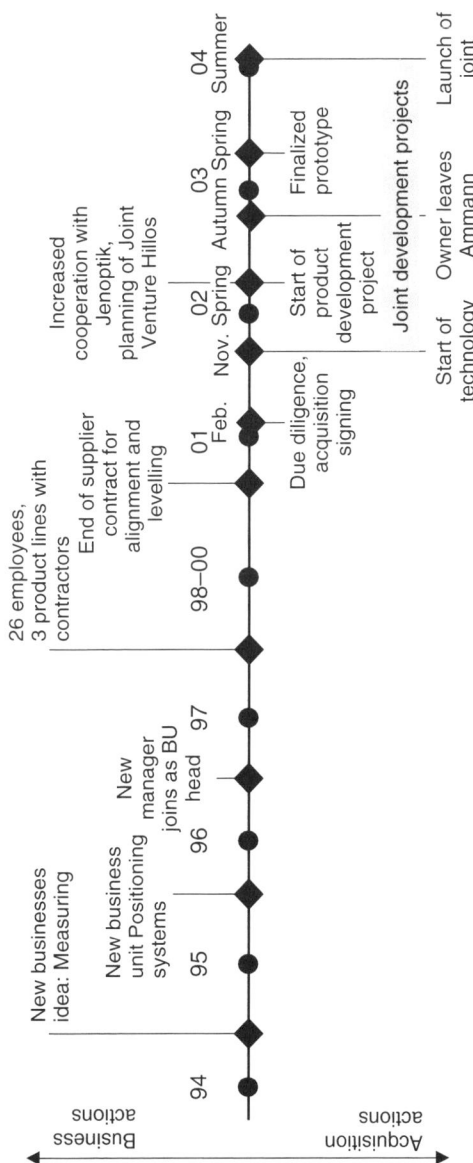

Figure 4.3 Acquisition timeline of Hilti–Ammann Lasertechnik acquisition

manufacturing and marketing added-value, top-quality products for professional customers in the construction and building maintenance industry. Hilti comprises several business units which are engaged in drilling and demolition, direct fastening, diamond, anchoring, firestop, foam, installation, screw fastening, cutting and sanding systems.

Hilti is characterized by its direct sales force of about 7,000 employees, its open and very strong company culture and its innovativeness. In 2000 the company employed 13,780 people and had revenue of CHF 3,138 m, of which it invested about 3.4 per cent in its R&D activities.

In 1994 as part of its continuous strategic development, Hilti was searching for new business fields. Thus a workforce analysis was conducted and came to an interesting conclusion (see Figure 4.4). It was shown that the core business of Hilti, which was drilling, amounted to a mere 17 per cent of the working activities and time spent by the main customers. The measuring part of the business, which was technologically not advanced and not flooded with competitors, amounted to 28 per cent of the working time. Thus Hilti considered the option of entering the measuring and positioning fields. At that time it had already developed one product in the field which served the detection of concrete reinforcements; however, it was part of the drilling business unit and treated more as an accessory than a product line. Three Hilti employees, one product manager, the head of marketing within the drilling division and an engineer diligently developed a business plan for entering the positioning business and introduced it to the board, which agreed to the proposed plan and thus the positioning business unit was formed in 1995.

As the members of the newly formed unit were quite inexperienced with the required products and technologies in the positioning and measuring

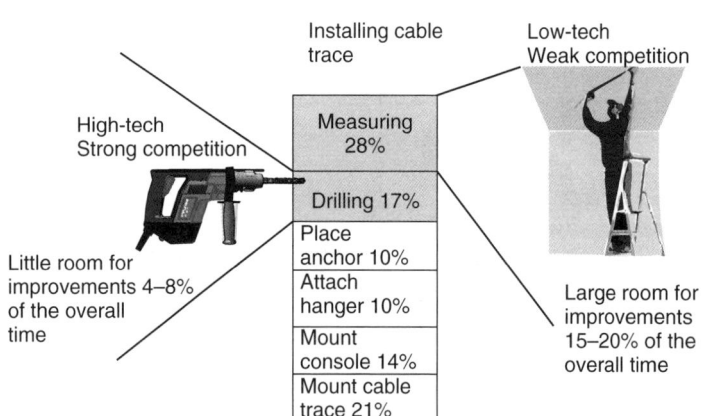

Figure 4.4 Potential for a new business

business, these competencies had to be built up. Thus as a first step a new head of the business unit was sought. A head-hunter proposed an experienced manager who was with Leica in the area of laser systems. He joined Hilti as head of the positioning business division in 1996. The new management team of the small division started to develop a strategy for its future activities and developments. Two different options were considered: Hilti could start developing, engineering and manufacturing positioning and measuring systems from scratch or it could initially cooperate with existing market players and slowly internalize the competencies to develop their own products later on. Whereas the first option was associated with high risks, the latter was characterized by a moderate profitability. However, due to the size of the unit and its lack of experience it was decided to first buy the products from the existing market players and then slowly internalize the required competencies. The objective was to launch internally developed products by 2003 or 2004.

Until 1997, Hilti's positioning business comprised three product lines:

- Levelling and alignment: products to support 3-D positioning via rotating lasers;
- Distance measuring: products for measuring distances;
- Detection devices: products already developed within the drilling business unit to detect concrete reinforcements.

Whereas the detection devices were made in-house, the distance measuring products were initially co-developed with Leica. However, when Leica cancelled its contract with Hilti, Hilti entered a cooperation with Jenoptik on the development of distance-measuring products. The levelling and alignment rotating lasers were delivered by Topcon, a large Japanese competitor. However, as Topcon, pushed by its distribution network, could not strengthen its relationship with Hilti, the cooperation regarding the rotating lasers was stopped. Subsequently, strengthening cooperation with Jenoptik, also in the area of rotating lasers, was considered; however, as Jenoptik did not have specific development competencies within this field, other options were investigated. At that time, in 2000, the positioning unit had twenty-six employees and needed to initiate its own development of products to keep on track with the pursued strategy to launch self-made positioning products. However, the group was still lacking the development competence to fully design and engineer a next-generation rotating laser. Thus the idea to acquire this competence externally was introduced. Soon the focus was on the owner of Ammann Lasertechnik because he excelled through his smart and technologically elegantly solved laser products for the construction businesses. His experience and capabilities were the development competence that was missing at Hilti.

4.3.1.2 Ammann Lasertechnik

Ammann Lasertechnik, based in Amriswil, Switzerland, developed and manufactured construction lasers with a high level of innovation and up-to-date

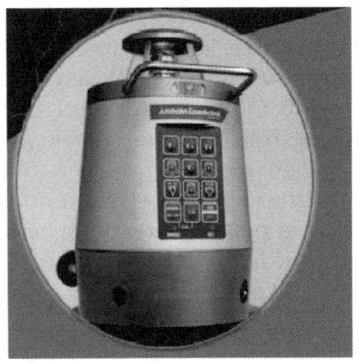 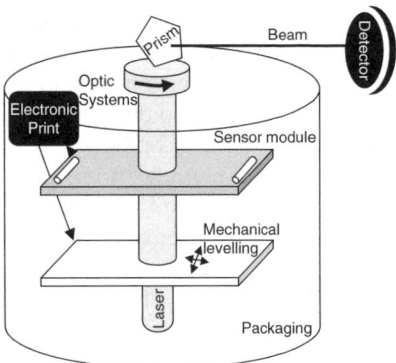

Figure 4.5 Rotating laser from Ammann Lasertechnik used for positioning and alignments

technology. The company was dominated by its owner, who not only owned the company but also excelled at engineering the products. The company employed around ten people and had a revenue of CHF 6 m in 2001. It produced laser systems, such as rotating lasers (see Figure 4.5), pipe lasers and grade lasers, used within the construction and building industries for alignment, measuring and positioning purposes. Ammann Lasertechnik's owner saw the opportunity to gain access to Hilti's huge distribution network and was enthusiastic about bringing in his development competencies. Furthermore, he was willing to step back from his daily business to focus on non-business-related aspects in life. Thus he was willing to sell the company.

4.3.1.3 The transaction phase

The division head of positioning systems and his colleagues developed a business plan to acquire Ammann Lasertechnik (LT) and to jointly develop the next generation of rotating lasers. The business plan, supported by financial figures and mainly measures to ensure the consistent integration of Ammann LT's owner's knowledge within the product development, was proposed to the board, which again agreed to this strategic path.

The due diligence was conducted very briefly. Only the financial aspects and IP issues were investigated. A detailed technological due diligence did not seem appropriate as the division head knew the owner's competencies and these were the only resources to be integrated and transferred to create value.

In February 2001 the acquisition was finished and the integration phase started immediately.

4.3.1.4 The integration phase

Hilti positioning systems pursued a very distant integration approach with Ammann LT (see Figure 4.6). The existing Ammann business was kept

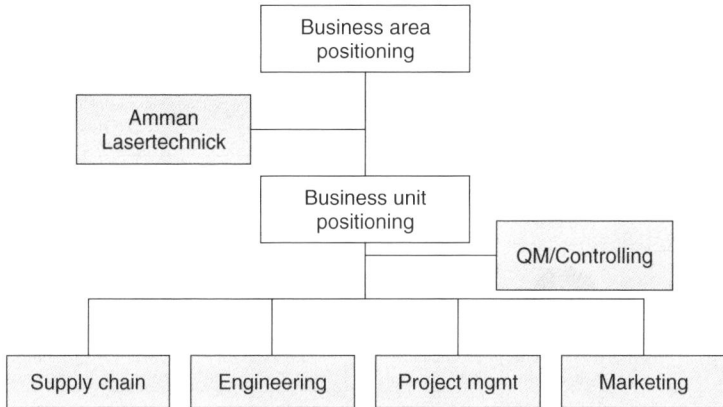

Figure 4.6 Distant organizational integration of Amman Lasertechnik

untouched, except that the owner was employed as a consultant for further product developments. Thus shortly after the acquisition, in November 2001, a technology development project started. In accordance with the lessons from the technology and innovation management, technology and product development projects are managed separately and have an assynchronal interface. Thus the product development is not delayed by immature technologies.

The technology development project was concerned with developing a new underlying concept and platform for the next generation rotating laser. The project aimed to conclude with proof of a concept to initiate the product development.

The technology development project was done by three people, the Ammann LT owner and two other employees from Hilti who brought in application experience from the Hilti side. The requirements, such as automatic levelling, measuring in horizontal and vertical planes, etc., for the new product were defined. The project was characterized by good and intense collaboration between the three people. The owner brought in his competence and largely determined the solution concept for the rotating laser, which became a smart platform product.

As part of the technology development project it was recognized that the acquisition of Ammann could provide only development competencies and not the manufacturing competencies which would be required to launch the product on a large scale. Thus Hilti decided to partner with Jenoptik with its extensive manufacturing and development know-how, albeit not in the area of rotating lasers. As a consequence it was decided that Hilti would utilize the joint venture with Jenoptik, called Hillos, which would take over the manufacturing of the alignment and levelling as well as Hilti's distance

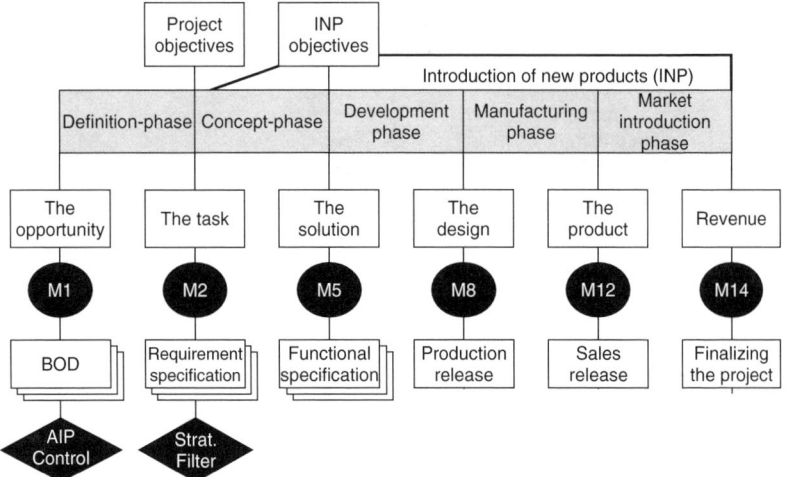

Figure 4.7 Time-to-market process at Hilti

measuring devices. Furthermore, the engineering team at Jenoptik would become a third engineering partner as it was experienced in developing mass manufacturing.

Thus in 2002 the product development process, following the implications from Hilti's TTM (time-to-market) process (see Figure 4.7) was initiated as a cooperation between Hilti, Ammann LT, Jenoptik and the joint venture Hillos. Further requirements, such as cost-efficient engineering tailored to mass manufacturing, were added. Each partner brought in specific competencies such as electronics, application engineering and packaging competencies from Hilti, design and optics competencies from the owner of Ammann LT, and additional engineering and manufacturing competence through Jenoptik and Hillos. The project team thus consisted of around nine people from engineering to marketing, managed and coordinated by a project manager. The engineering teams were organized according to modular structure such as system engineering, construction and electronics. The project coordination was mainly done via video conferences and meetings. Every week one meeting within Hilti ensured continuous coordination. Jenoptik and Hilti team members met only once every six weeks to discuss the interfaces between the various models. The Ammann LT owner participated only sporadically. Whereas some modules were co-developed by the former owner and Hilti people and the cooperation was working quite well, the engineering of modules together with Jenoptik was much more difficult. The owner had only roughly sketched his concepts developed within the technology development projects. These were transferred to Jenoptik, which had not participated in the technology development project. Subsequently, the purpose and elegance of the concepts were misunderstood, wrongly interpreted and finally changed, which led to disagreements between the former head of Ammann LT and other team members and delayed the product development process. Furthermore, the former owner, not accustomed to high-volume product considerations, complained about the enduring and partly inflexible product development process imposed by Hilti. Whereas product development at Ammann LT took around half a year, this project lasted over one-and-a-half years. This resulted in the owner withdrawing from the project centre. He had the feeling that the engineering had been taken over by Jenoptik and his input was insufficiently accepted. Thus his engagement at certain times was reduced and required occasional remotivation.

However, the jointly developed product is now, in the beginning of 2004, due to be launched and will most probably be a success. Despite this successful integration of the engineering competence of the owner of Ammann LT, it could not be transferred and retained in the long run. Even though Hilti intends to take advantage of his consultancy work, the former owner will be less likely to continue in this way. One negative side-effect of this acquisition was the complete loss of innovativeness for the Ammann Lasertechnik company. As the owner was no longer the manager, product development was no longer pushed and the company is now losing competitiveness. To oppose this threat, Ammann LT will use the Hilti-developed product platform for future products.

Today Ammann LT continues its operations at a distance from the Hilti operations and partly lacks professional leadership.

4.3.1.5 Interrelations, conclusions and discussions

Hilti's acquisition of Ammann Lasertechnik can be characterized as a friendly, small acquisition of new technologies within a medium-paced market. The objective of the acquisition was to integrate and internalize product development technologies for the levelling and alignment product line. This objective was partly achieved. The acquired competence was used to develop a new product platform. Nevertheless, this competence could not be fully internalized, as the former owner of Ammann LT withdrew at a certain level from the product development projects, his competencies could not be fully learned and as a result these had to be compensated for with Hilti internal efforts. Thus the acquisition created short-term value only.

Now the question arises of how the achieved and unachieved technology-based value-creation opportunities were dependent on the initial conditions of the acquisition, the external developments and the strategy processes.

Generally the *initial conditions* were quite favourable to technology-based value creation. Despite Hilti's lack of acquisition competencies, as acquisitions are generally not a strategic focus of Hilti, the division head brought some acquisition and integration experience with him and thus knew how to manage the process. Furthermore, Hilti pursued a very smart way of building up a new business. Whereas other companies would have immediately acquired a company to enter a new business, Hilti first built up internal competencies with the hiring of the new division head and the cooperation projects and subsequently acquired a company. This provided the business unit with sufficient absorptive capacity to identify, use and internalize competencies within the rotating laser business. Furthermore, the acquisition was one logical step in pursuing the business unit strategy and thus fitted very well from a strategic perspective. Hilti was aware that Ammann LT could only be a partner regarding the product development and therefore, in consequence, an additional manufacturing partnership had to be established.

The technology-based value creation potential was relatively attractive as the owner mastered the key technologies very well. However, his abilities and experiences were highly tacit and thus difficult to leverage. Thus although the integration of the owner's knowledge within modules of the product was quite easy to achieve, the transfer of his capabilities was hindered by their tacit character. Additionally, the contextual differences, which became apparent from the former Ammann LT head's complaints regarding the formalization of the TTM process, hindered open communication and collaboration.

The *acquisition process* did not impact the potential for value creation, especially as the investigations were quite easy to conduct mainly due to the small size of Ammann Lasertechnik and the division head's knowledge about the owner's competencies. The *integration process* has focused on the retention and integration of the former owner and intentionally kept the Ammann Lasertechnik business completely separate. Thus Hilti was dependent on the owner's presence. When he backed out of the project the competencies were no longer at Hilti's disposal. Thus this integration approach was conducive to immediate cooperation and product development projects but did not result in a long-term cooperation and knowledge transfer. The *external developments*, such as market developments or technological advances, were not mentioned as having a significant impact on technology-based value creation.

4.3.2 Case 2: Fillpack–Aseptofill acquisition

In January 1999 Fillpack Holding AG acquired German-based Aseptofill Maschinenfabrik for DEM 30 m. 1998 Aseptofill, the manufacturer of aseptic filling machines for PET bottles and cups, reported net sales of DEM 57 m with around 400 employees. The acquirer aimed to source competencies in aseptic filling for plastic bottles (for a detailed timeline of the acquisition see Figure 4.8) and Aseptofill was integrated into the Fillpack Plastics division. The acquisition was described as only partially successful by Fillpack.[6] The main reasons for the failure were rooted in technical and integration problems.

4.3.2.1 Initial situation

Fillpack Holding AG was founded in 1840 in Germany, and from then on diversified steadily over the decades to produce a wide range of machinery

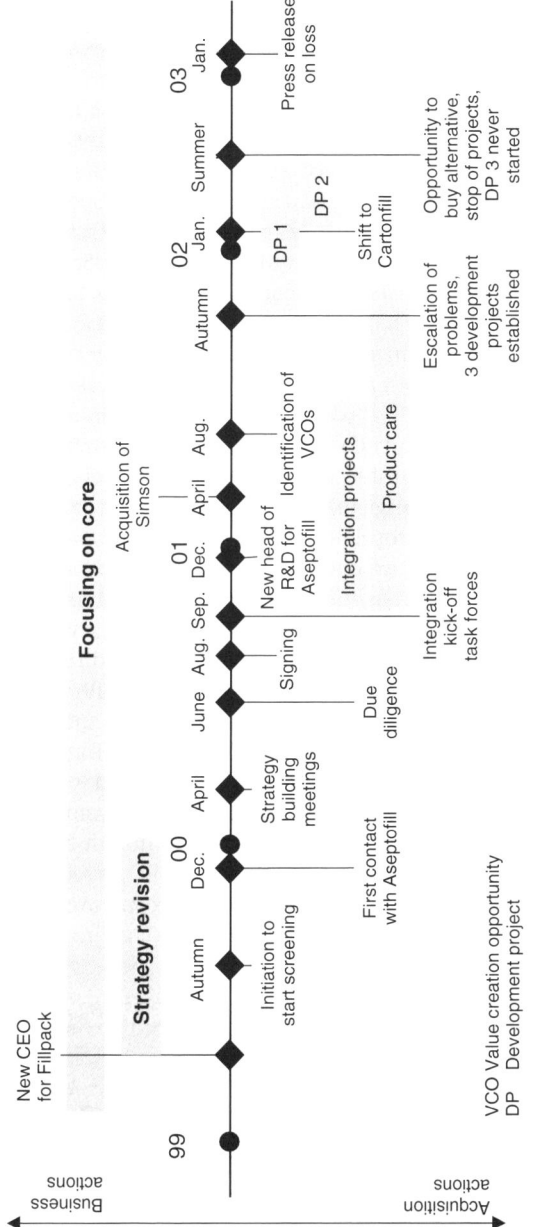

Figure 4.8 Acquisition timeline of Fillpack–Aseptofill acquisition

for automation and packaging equipment. In 1998 Fillpack was organized into two divisions: Fillpack Cartonpack, No. 4 supplier of machinery and cartons to processors of beverages and liquid foodstuff; and Fillpack Pack, division for manufactured packaging systems and machines to service the food, health and body care industries.

At that time Cartonpack was the most profitable division and responsible for the largest revenue share. However, it was threatened by a slowdown in market growth and the emergence of substitute packaging technologies, such as HDPE and in particular PET packaging, for liquid food. Furthermore, the company's CEO who had been in his role since only October 1996 had unexpectedly announced his resignation in November 1998.

Thus the board members of Fillpack, facing a highly uncertain financial perspective for Fillpack's future, decided to assign a new CEO to redirect and revive the company. His appointment took effect in May 1998. He addressed the upcoming challenges posed by the rising PET market (alone 10 per cent annual growth since 1990). PET packaging is highly attractive due to its transparency, formability and its flashy appearance. The main disadvantage of PET so far is that the shelf-life of carbonated and oxygen- and light-sensitive products, such as orange juice, which includes vitamin C, is largely limited due to the lack of appropriate barriers within PET bottles.

The first inroad of Fillpack into beverage filling using plastics was through an acquisition. With this acquisition Fillpack acquired competencies in extrusion blow-moulding for HDPE bottles and in PET bottle manufacturing. The whole acquisition became part of a new division called Fillpack Plastics.

Nonetheless, with the newly acquired competencies Fillpack still could not compete with the strongest market players Krones[7] and Sidel[8] in the plastic packaging market for liquid food, especially as the filling competence for plastic bottles was lacking (see Figure 4.9). Instead of focusing on the slow-growing non-aseptic beverage market, Fillpack wanted to address the upcoming aseptic filling of plastic bottles, even though the related barrier technologies were not yet ready. Therefore, the CEO gave rise to the idea of acquiring this filling competence and asked the corporate M&A team to

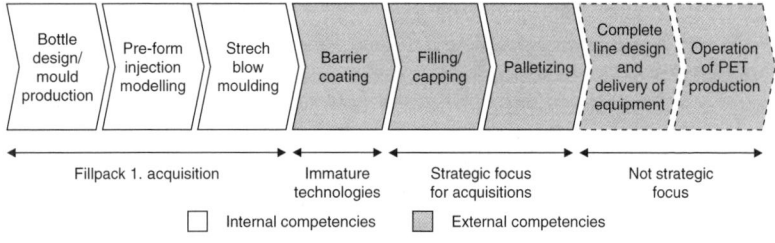

Figure 4.9 Strategic competencies in the value chain of aseptic filling of plastic bottles

screen for suitable candidates. The option to build a filling machine for plastic bottles in-house was not seriously considered, especially as the employees, so far focusing on cartons, saw PET as a competitor and not as an interesting new internal business.

4.3.2.2 Corporate M&A at Fillpack

Fillpack's corporate M&A team directly reports to the executive board of the Fillpack Holding. It is in charge of setting guidelines and providing mechanisms to support the acquisition process. Furthermore, the team has to ensure the quality of the M&A process at both group and business unit level, to build and maintain an understanding of markets and candidates at corporate level and to provide M&A support to the functional M&A teams.

The basis of the acquisition management is the acquisition process initiated by the business strategy, mainly consisting of five phases which run either partly sequentially or in parallel. The process is initiated by the corporate and business unit strategy, which is co-developed by the corporate M&A team. The tasks and roles within the acquisition process are distributed between the Corporate M&A team and the business unit. Once an acquisition is initiated, an acquisition organization consisting of a steering committee and functional teams is established. The size and staffing depend on the size of the acquisitions. The steering committee for larger projects consists of the CEO, CFO, Head of Corporate M&A and board members or business unit heads as needed.

The typical due diligence areas cover fields such as accounting, corporate structure, environment, finance, human resources, insurance, legal, operation and manufacturing, organization and management, strategy and R&D. The R&D due diligence is concerned with new product developments, new market introductions and key research efforts of the target company. The due diligence consists of the bid preparation, the bid process, the due diligence preparation, the document request list, the data room evaluations, the management interviews, site visits and the final price negotiation

The corporate M&A team, which consisted of three people, was moderately experienced in carrying out acquisitions. Despite the average acquisition success rate no acquisition and integration controlling instrument had been established. It was explained that due to the open culture within Fillpack, feedback and thus learning would come inherently.

4.3.2.3 Screening for the candidate

To support the corporate M&A team's screening activities in identifying a company with aseptic filling competencies for plastic packaging, McKinsey was mandated to do a survey of clients and to obtain information regarding potential candidates. Furthermore, trade fairs, the internet, talks with experts and other sources of information were consulted to identify a potential target. A few companies such as Serac, Simson, GEA and Aseptofill repeatedly appeared

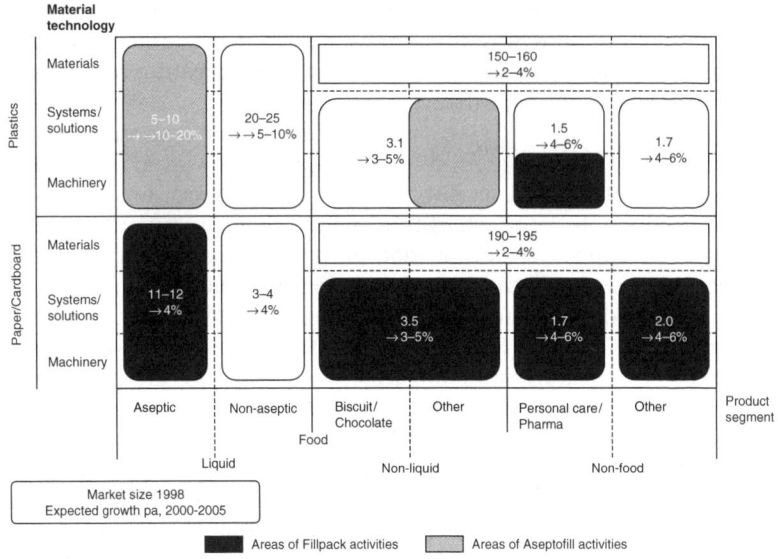

Figure 4.10 Aseptofill activities match the strategic focus of Fillpack

in the discussions. Thus talks began with the companies and it soon turned out that Aseptofill, with the aseptic filling machines for plastic bottles, would be a good match (see Figure 4.10). The CEO personally initiated the talks with the owner of Aseptofill.

4.3.2.4 The Aseptofill company

Aseptofill was established in 1939 in Germany and focused on the engineering, manufacturing and selling of filling machines for preformed cups. Aseptofill's traditional business since has been the filling of yogurt, ice-cream and frozen desserts. As this market was quite limited, the owner decided by the end of 1990 to develop a fully aseptic filling machine for plastic bottles. The first machine was delivered to Brazil, and machines were still being produced or ordered in 2000. At that time, Aseptofill had about 400 employees, revenue of 28 m and 0.6 m of margin in 2000, with a deteriorating development, and annually invested approximately 3 per cent of its revenue in R&D. The whole organization was highly influenced by the patriarchal management style of the owner. The company had no R&D organization, the business was entirely order driven and managed on an *ad hoc* basis. At weekly meetings the different orders were discussed and tasks were assigned. The employee's knowledge was highly dispersed and tacit in the company.

4.3.2.5 The transaction phase

Soon, in December 1999, the acquisition intention of Fillpack became quite concrete and the company sent two engineers from the Fillpack Cartonpack division to Aseptofill to investigate the company and its technologies. These engineers had competencies in aseptic filling, but only from carton packages. This preliminary assessment was followed by a more detailed financial investigation by an external specialist in April 2000 and a detailed due diligence in June 2000.

The technology due diligence was concerned with the analysis of the existing machines and especially the newly developed aseptic filling machine. The first machine installed in Brazil seemed to have had some starting difficulties; however, this was due not to the technologies from Aseptofill but the application of the customer. Furthermore, it was discovered that the technologies and modules were generally well mastered and worked well, but some aspects of the sterilization technologies had potential for improvement. As the investigators from Fillpack had no experience in bottle-handling, they compared the Aseptofill machines with another aseptic plastic filling machine for plastic bottles, the Asbofill, developed by Aseptofill's competitor. The technology due diligence was thus characterized by the investigation of the machines and their engineering. The next-generation products and technology integration within the whole Fillpack was not a critical factor. During conversations it became clear that the new aseptic filling machine was not the core competence of Aseptofill and that its focus was still on the traditional filling machines for yogurts and the like where aseptic conditions were not as critical.

Other aspects which were given attention during the due diligence were the cultural differences between the two companies and the dominant leadership style of the owner. These differences and the owner's willingness to stay with the company were identified as potential threats to the acquisition integration.

The company assessment was based on three different scenarios and concluded that Fillpack and Aseptofill jointly had large potential for synergies and thus would give the consolidated company value. The main value-creation opportunities were seen in an improvement of the market position of Fillpack in the PET market through controlling the whole filling process and competencies and, furthermore, in consolidating the sales and administrative functions and in using the Fillpack know-how to decrease the working capital requirement. These synergies, which were mainly based on the successful innovation and resource-deployment opportunities of the joint resource base of Aseptofill and Fillpack, added up to a potential value creation of CHF 28 m (see Figure 4.11).

Without paying more attention to the following integration phase, the proposal to acquire Aseptofill was submitted to the board of Fillpack.

	Driver	Assumed Potential	Increase in Value
Market synergies in PET business	Improve market position in PET market through controlling the whole filling process & competencies	Achieve 20% in market share in aseptic PET filling	17
Market synergies in dairy business	Leverage position of Fillpack cartonpack in emerging markets (aseptic techn)	Increase sales volume in emerging markets	0
Cost synergies in dairy food and PET business	Consolidate sales and administrative functions	Cost cutting in service sales, distribution, administration, procurement	9
Synergies through net working capital	Use the know-how of Fillpack to decrease the working capital need	Reduction of stock and cycle time	2
		Total	28

Figure 4.11 Value-creation opportunities pursued through the Aseptofill acquisition

On 31 August 2000 it was announced that Fillpack had taken over Aseptofill on 1 July 2000.

4.3.2.6 The integration process

The integration process was initiated by a kick-off meeting at the Aseptofill headquarters. The Fillpack CEO introduced Fillpack and its strategy and outlined the pursued synergies and upcoming changes. Furthermore, he stated that Aseptofill would become part of the Fillpack Plastic division. The Aseptofill personnel, even though they were insecure about the future, generally welcomed the new ownership of Fillpack, especially as the dominant leadership style of the former Aseptofill owner had not been easy to cope with for some employees.

The integration process was initially managed by task forces, which were established by the corporate M&A team (see Figure 4.12). These task forces were concerned with the realization of potential synergies and some supportive integration projects such as the introduction of an ERP and a CAD system. It was agreed that the former owner, due to his profound knowledge of the business, would stay another year with the company, taking over operational and R&D-related tasks. Additionally, a new head was assigned to Aseptofill. He was a friendly, cooperative and helpful type of person rather than a tough straightforward business manager. He was in charge of keeping the Aseptofill business running beside the integration projects and of taking over the overall integration lead after the task forces had finished.

The task force teams combined the sales forces and tried to transfer quality standards to the Aseptofill engineering and manufacturing areas. Furthermore, CAD systems were introduced to Aseptofill engineering quite quickly, which led to the departure of some engineers. These integration

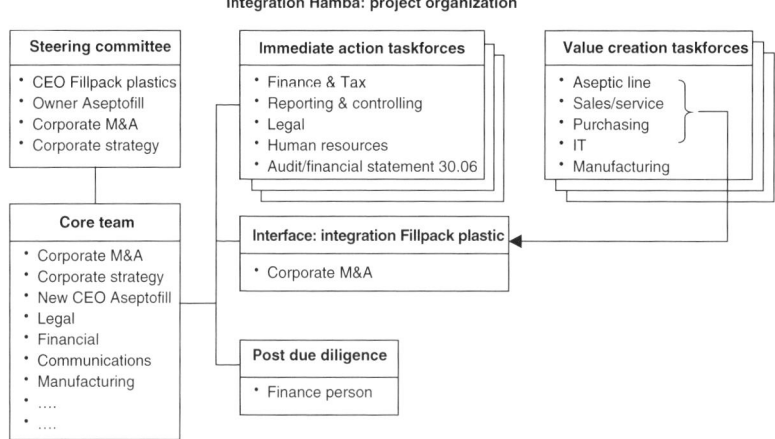

Figure 4.12 Integration teams at the Aseptofill acquisition

projects were highly resource- and time-consuming and resulted in neglect of the daily business at Aseptofill.

Additionally, a new building for Aseptofill near to the original facilities was planned – a project which also consumed the attention and availability of the Aseptofill management and employees.

In parallel to the large integration effort, technological difficulties appeared. It turned out that the machine delivered to Brazil had more technological problems than expected and the customers started to make claims on their warranties. Reacting to this threat, a new R&D manager was introduced to Aseptofill in December 2000. He initiated projects to improve the already-delivered and currently developed and constructed machines. As part of this product care project initial knowledge transfer between Fillpack and Aseptofill was achieved.

Furthermore, the new R&D manager, and some other engineers also with the CTO, identified technology-based value-creation opportunities in terms of possibilities of transferring existing knowledge and technologies mainly from Fillpack Cartonpack on filling technologies, sterilization and sealing technologies to Aseptofill. However, these projects, defined in August 2001, never really started as Aseptofill's R&D was completely absorbed in getting the machine running appropriately rather than improving it.

In parallel, a new order from a large customer was received, and while constructing this new machine additional technical difficulties appeared mainly due to different plastic bottle forms. Thus at the end of 2001 Aseptofill faced large technical problems which needed to be solved, technological synergies which needed to be achieved (see Figure 4.13) and deadlines to be met for machines that had already been sold. In this situation,

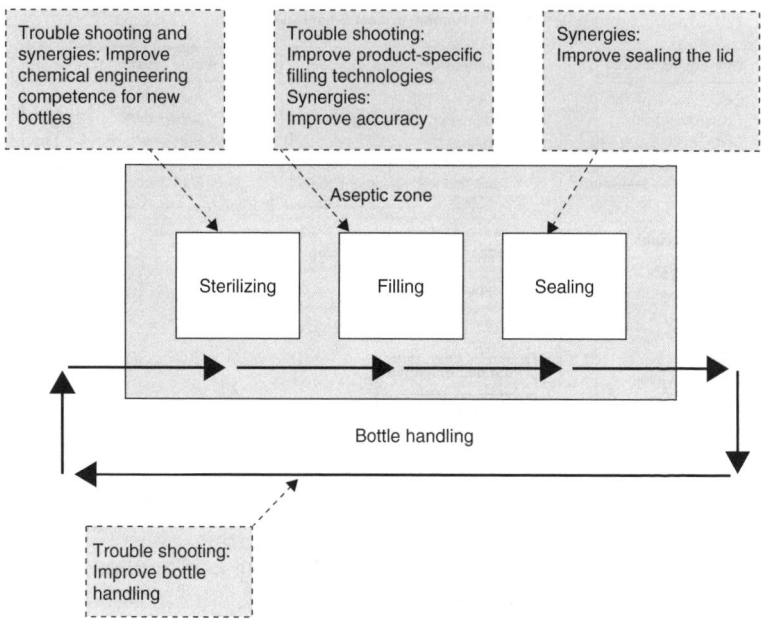

Figure 4.13 Trouble shooting and technological synergies at the Aseptofill filling machine

the corporate management noticed the difficulty Aseptofill was in and ordered an immediate stop on sales and manufacturing in order to provide time to redesign and re-engineer the product. To some extent Fillpack was lucky that the PET market and especially the aseptic filling of PET bottles had not developed as fast as previously expected. This was due to the general economic downturn and difficulties in developing an appropriate PET barrier required for the long-term storage of carbonated or vitamin C-containing liquids in PET bottles.

As corporate M&A only kept track of the financial aspects of the integration, it was unaware of Aseptofill's operational and technological problems. Furthermore, Fillpack had acquired another much larger manufacturer of plastic bottle-filling machines, Simson, which in 2001 had consumed the more strategically oriented attention of the corporate M&A.

As a reaction to the problem, three new cross-divisional teams were established (see Figure 4.14). The first team, a combination of Aseptofill and Simson engineers, were given the task of improving Aseptofill's filling technologies. The second team, comprising Cartonpack and Aseptofill employees, was to improve the sterilization function of the Aseptofill machines and the third team, staffed with people from Cartonpack, Simson and Aseptofill, was

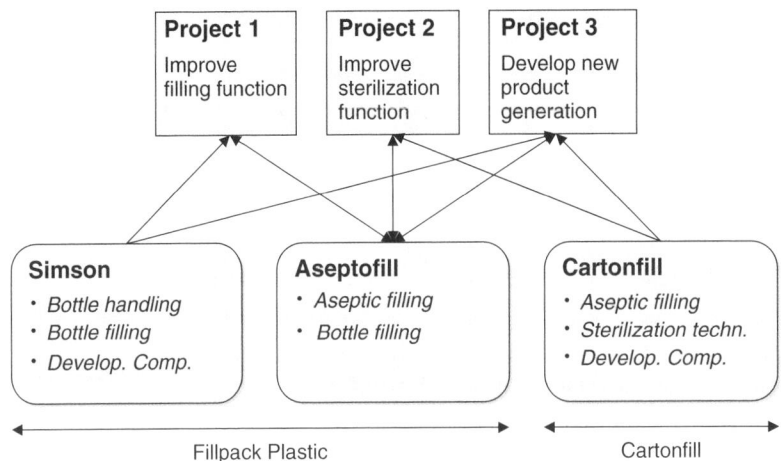

Figure 4.14 Joint development projects to transfer technologies and develop new products

in charge of developing a completely new product generation with a modular structure. The estimated duration of this project was approximately one year.

As Aseptofill at that time was part of the same division as Simson, the cooperation was not difficult to achieve. The project between Cartonpack and Aseptofill faced greater difficulties mainly because of cultural differences. Knowledge transfer was more on a voluntary base and thus Aseptofill was shifted under the responsibility of Cartonpack, which gave rise to good cooperation and knowledge transfer between the two business units.

In mid-2002 the opportunity arose to buy the Asbofill filling machines, a competitor to Aseptofill's filler. It was decided that instead of investing their own engineering capacity of all three business units, Aseptofill, Cartonpack and Plastics, in re-engineering the Aseptofill machine, buying a new machine would be more appropriate. It was concluded that Aseptofill would return to its original and profitable core business of non-aseptic filling machines for yogurt and the like. A turnaround manager was assigned to bring Aseptofill back to profitability.

A press release (see Figure 4.15) at the beginning of 2003 made a public statement regarding the acquisition and its result. Additionally, some of the workforce was dismissed in order to reduce the production depth. The company is now (Autumn 2003) structured under the division 'Other', indicating that the operations are no longer a core business of Fillpack.

One major action following this acquisition was the foundation of FillpackTec, a corporation that owns the technologies of all businesses within Fillpack. This centralization of a technology pool and matching coordination by the technology management team at Fillpack is designed to

> Fillpack press release, 28 January 2003
>
> ....
>
> Losses in the three companies [acquisitions] Aseptofill, Fill X and Fill Y, however, had a detrimental effect on earnings. In addition, goodwill adjustments at Aseptofill and higher taxes took their toll. As a result, the group expects to post a net profit of about EUR 9 million.

Figure 4.15 Fillpack Holding's press release, 28 January 2003

support cross-divisional cooperation and thus to achieve inter-business unit technology-based value-creation opportunities.

4.3.2.7 Interrelations, conclusions and discussions

The Aseptofill acquisition conducted by Fillpack can be characterized as a relatively small friendly acquisition[9] of a company with related customers but a new and unfamiliar competence for Fillpack in a fairly dynamic market.[10] The acquirer aimed to incorporate the competencies and to reap synergies through integrating these within the upcoming product lines. However, the pursued, planned and partly emergent technology-based value-creation opportunities could not generate sufficient value.

The *initial conditions* of the acquisition were partly favourable to and partly negatively impacting on technology integration and product developments. On the one hand Fillpack had moderate acquisition experience and a formalized M&A management and the acquisition fitted very well from the strategic perspective. On the other hand there was a lack of technology-based value creation potential. As it turned out in the course of the integration, the potential to jointly create value from products and technologies was quite small compared to previous expectations. This was partly caused by Aseptofill's poor level of mastering its own technologies. This became obvious from the large technical problems of the already-delivered machines. Furthermore, as a result of the non-modular structure of the aseptic filling machine, the underlying technologies were difficult to combine. Last but not least the aseptic filling technology for PET bottles in general turned out to be less attractive than expected, mainly due to its dependence on appropriate PET barriers. As these barrier technologies were still immature and very expensive, the market for aseptic filling for PET bottles was still limited. Thus the effective technology-based value-creation opportunities were more costly to achieve and not as attractive as initially supposed. Additionally, Fillpack lacked sufficient background knowledge to appropriately investigate the new competencies in the aseptic filling of PET bottles. Furthermore, the specific human resources and management attention required for integration were insufficiently available, especially as subsequent larger acquisitions were conducted shortly after the Aseptofill contract was signed.

Another factor hindering technology-based value creation was the different corporate contexts of Aseptofill and Fillpack. As was already noticed during the due diligence, the cultural, organizational and business logic characteristics greatly varied between the two companies. Aseptofill as a family-owned company was highly dominated by the patriarchal leadership style of the owner. Furthermore, the organization and business logic were completely tailored towards project business and existing customer contacts. The characteristics of Fillpack as a large, professional-thinking manufacturing business were very different. Fillpack is attributed with its quite open communication and direct leadership style, a functional organization within the different businesses and a business logic tailored towards shareholder value and joint value creation. These contextual differences between the two companies hindered communication, interaction and thus joint value creation. Statements such as 'we were talking in different languages' often occurred during the interviews.

The *strategy processes* were very ambiguously managed. The acquisition process was generally managed very well. The only pitfall was the shortcomings within the technology due diligence. Technological maturity, attractiveness and combination ability were overestimated or poorly considered. The CTO also remarked that he was highly involved in strategy building and partly also the integration phase, but during the transaction phase his involvement was quite limited. The integration was much less consistently managed. In the beginning of the integration process the resources and people were entirely focused on value creation and supportive projects, resulting in neglect of sustaining the existing business. Furthermore, the abrupt transition from acquisition to integration process caused a lack of focus. The new Aseptofill CEO was introduced to the company when the transaction and especially the whole strategy-making process were already over. Furthermore, the support from the corporate M&A was limited in the integration phase, as the team was already concerned with the next acquisition. Thus the integration was not actively managed as a top-priority project but became part of the daily business. Furthermore, it was reported that the leadership style of the new CEO did not fit with the integration task, where a strict and direct leadership style would have been more suitable.

The *external developments*, which were mainly the slower-than-expected rise in the aseptic filling of PET bottles, were somehow positive for Fillpack, as it was entirely covered by integration efforts and the product care for the already-delivered products and thus a strong market growth would have overextended Aseptofill's abilities at that time.

4.3.3 Case 3: Phonak–Unitron acquisition

The Swiss Phonak Group, the fifth largest hearing-instrument manufacturer in 2000, on 6 November 2000 announced the acquisition of the world's seventh largest, Canadian-based hearing-instrument manufacturer, Unitron Industries Ltd, for CHF 161 m. The deal was completed on 1 January 2001. Phonak aimed to use this acquisition to gain access to the mid-range market segment and to expand its geographic scope. The beginning of the integration of Unitron was marked by unexpected difficulties mainly due to an overestimation of the technological competencies of Unitron's suppliers and problems with the low integration approach (for a timeline of the acquisition see Figure 4.16). In the course of the integration, operational and technology-based value-creation opportunities became more important and also necessary to sustain the market dynamics. Thus this case shows the emergence of technology-based value-creation opportunities and outlines the lessons learnt from their results. Even though, over the following years, technological synergies were achieved, the integration was marked by two press releases describing losses due to the acquisition and integration of Unitron of CHF 12 m and 75 million.

4.3.3.1 Initial situation

Phonak was founded in 1947 and since then has been in the hearing-aid business. It is a company driven by technology and innovation, investing around 6–7 per cent of its annual sales volume in its R&D unit. Before the acquisition in 2000, Phonak employed 1,279 people, of which 511 were in Switzerland. In fiscal year 99/00 the consolidated revenue was CHF 314 m, with an EBIT of CHF 49 m. Phonak Holding included two business segments, Phonak AG in Stäfa, Switzerland and Communications AG in Murten, also in Switzerland. Phonak covered the Swiss and European as well as the North American market with its primarily premium customer products, whereas its stake in Asia was very small. Phonak's product range mainly covers four different groups: full digital hearing computers (Claro line), digital programmable hearing systems (PiCS, E-PROM, Astro), classical hearing systems (Classica) and wireless communication (MicroLink, Watch Pilot).

Phonak competencies were in the digital and analogue technologies for hearing devices. It differentiated itself through extensive capabilities in designing and manufacturing the products, in developing the signal processing software and the application-specific integrated circuit (ASIC). Phonak has a tremendous manufacturing depth and know-how in acoustics, medicine, physiology, and so on.

In 1999 the hearing-aid business was in a turbulent phase. Not only was digital technology conquering the business but also a large industry consolidation was rapidly decreasing the number of large competitors within Phonak's business areas. Thus the Phonak management decided at the five-year-planning meeting in 2000 to increase the current market share of about 8–9 per cent up to 20 per cent also by means of acquisitions. To reach this market share Phonak could not stick to its traditional high-end market segment but had to acquire a company in the medium and lower segment.

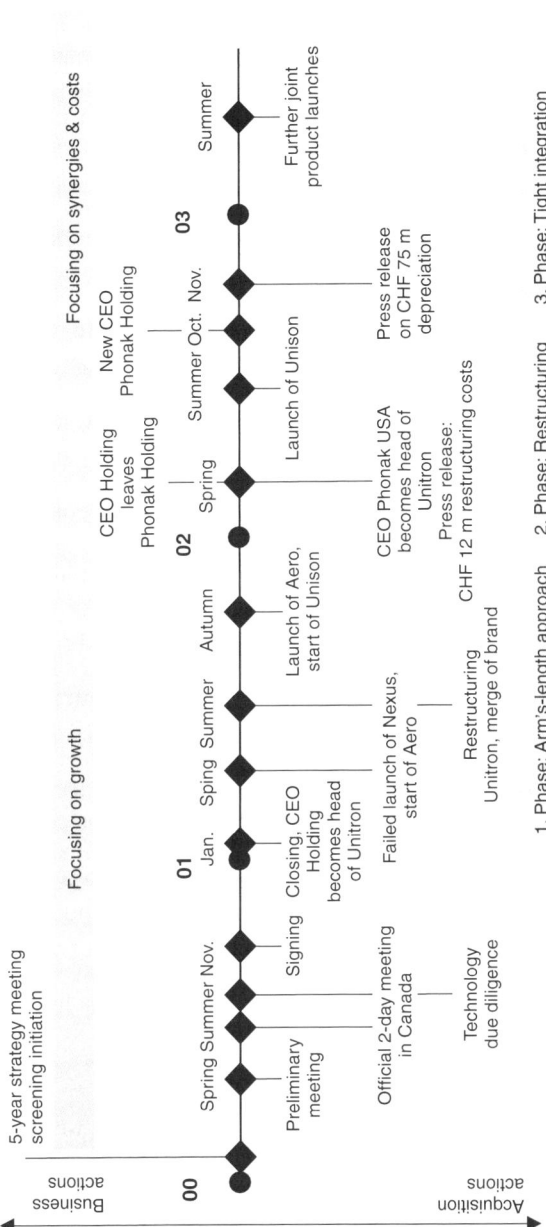

Figure 4.16 Acquisition timeline of Phonak's Unitron acquisition

Additionally, an increase in geographic scope such as Asia would have been a preferable match to the strategy agreed upon. Furthermore, another brand could provide Phonak with greater flexibility in serving the overall market. Thus at the end of the meeting each member of the management was assigned the task of searching for potential acquisition opportunities. As Phonak did not have a corporate M&A team, its acquisition and integration capabilities were quite limited and inexperienced.

A few months later the CEO of the very successful Phonak USA, Warrenville, Illinois (50 per cent of revenue is earned there) came up with the idea of acquiring the Canadian-based Unitron Company.

4.3.3.2 The Unitron company

The 40-year-old Unitron Hearing is headquartered in Kitchener, Ontario, Canada with operations in Canada, United States and Germany and a global network of distributors. Unitron itself had acquired two companies, Lori Medical Labs in August 1999, and Argosy in March 2000, which had not been integrated by the end of 2000. Unitron employed around 600 people and had revenue of around CAD 96 m (CHF 110 m) in 2000. Its worldwide market share reached about 3.4 per cent with a strong focus on America and also Japan.

Unitron had well-known and good branding as a stable though not highly innovative company in the medium segment, with a minor investment in R&D of about 3–4 per cent of its annual revenue. The company had a wide product range and a new fully digital product, called Nexus, in its pipeline which was designed to meet the demands of the upper segment. Production of Unitron products was highly outsourced, thus production depth was not very high. Unitron's core competencies were in the development and programming skills of an open fully programmable signal-processing platform (Toccata) for hearing instruments. This development competence was spun-out in a company called DSP factory, which allowed the marketing of the technology platform, also externally, from Unitron. In addition, process and product technologies could only partly live up to the profound capabilities of Phonak.

Unitron recognized that it could not sustain its business with its short-comings in technological competence and product pipeline in the long run and thus hired a new CFO to make the company public – either via an IPO or by being acquired. The owners of the company also agreed on this proposal.

The CEO of Phonak USA recognized this situation and a first preliminary meeting in Switzerland was organized in Spring 2000. The chairman and CEO of Phonak Holding as well as the CEO and CFO from Unitron participated in that meeting. The meeting was characterized with mutual respect and a friendly atmosphere. The CEO of Unitron summarized the situation by telling his CFO: 'we had better do something with them, because we are not going to beat them'.

4.3.3.3 The pre-transaction and transaction phase

The idea of acquiring Unitron Industries Ltd was positively accepted by Phonak's management. A second, two-day meeting took place in May 2000, where the CEO of Phonak Holding, the CEO of Phonak USA and the Head of R&D visited the company in Canada and in the US. They were informed in more detail about the company and its future developments. After their return, the Phonak CEO set up a project team consisting of himself, the USA CEO, Phonak's CFO and, on standby, the head of R&D and a senior expert engineer (see Figure 4.17). The project, however, was managed on a demand basis and information was shared mainly between the CEO of the Phonak Holding and the CEO of Phonak USA. The overall acquisition strategy was generally communicated with the following objectives:

- Develop a double branding business in order to approach all market segments;
- Address low segment and the Japanese market;
- Realize market synergies through scale in volume of procurement.

Additionally, the head of R&D and the senior expert engineer visited Unitron in Canada to conduct the technology due diligence; however, as there had been no clear indication on the technology strategy within the acquisition, the technology due diligence was organized more broadly. The team investigated the strengths and weaknesses as well as the opportunities and threats challenging Unitron. Furthermore, they identified several

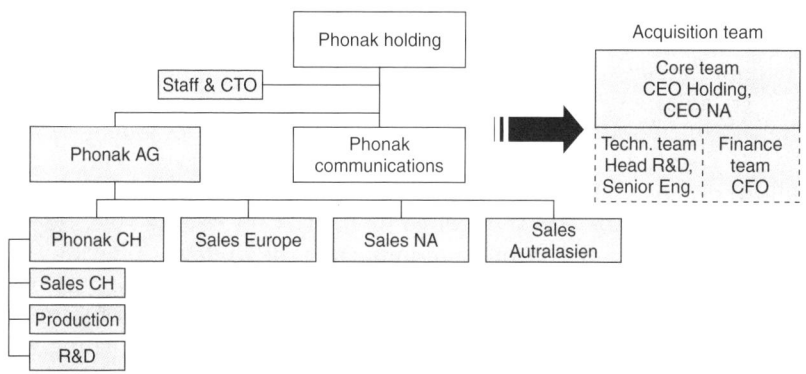

Figure 4.17 Acquisition team at the Unitron acquisition

value-creation opportunities such as the shared use of skills and knowledge in programming Toccata, the DSP factory, signal processor. They investigated the upcoming products and contracts with DSP factory, the provider of the Toccata platform. They came to the following conclusions:

- The upcoming Nexus product was almost finished;
- The underlying DSP factory Toccata technology might be a risk factor due to its immaturity;
- The product portfolio was highly diverse and contained many old products;
- The cost structure of the assembly line was not sufficiently under control;
- The packaging and construction technologies were not state-of-the-art.

Despite the caveat that some technologies might be immature and that technology-based value-creation opportunities can occur, these issues were paid significant attention neither at the signing of the acquisition contract nor at the integration planning. This planning was limited to the announcement that the Unitron Company would be integrated with an arm's-length approach and only a few internal integration projects would be launched. Furthermore, a slow and sensitive integration approach was to be pursued.

The signing was announced on 6 November 2000 and became effective on 1 January 2001.

4.3.3.4 The integration process

The integration process of the Unitron Company can be separated into three different phases (see also Figure 4.16):

- The arm's-length approach
- The restructuring phase
- The integration phase.

In a first step the CEO of the Phonak Holding, Phonak AG and Phonak Communications took over the CEO position at the Unitron Company as well. Organization-wise the Unitron Company became another division of Phonak Holding without any boundary spanning structures to Phonak AG, the sales organizations or Phonak Communications (see Figure 4.18). However, due to the CEO's limited time, the internal consolidation of Unitron was managed and supported by the CEO of Phonak USA. He initially focused on consolidating and improving the Unitron Company as it stood before starting to reap any operational synergies. Thus he improved cost transparency in the assembly lines and introduced task force teams to implement an existing ERP system and to consolidate the purchasing volume. Furthermore, he moved the two acquired companies in the US into one new building and merged the three brands into one new one: Unitron Hearing. These integration projects were conducted very quickly and in a

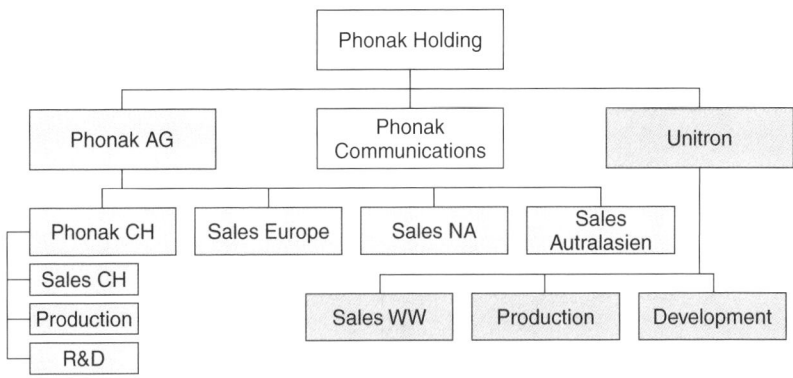

Figure 4.18 Initial organizational integration of Unitron

straightforward manner which caused on the one hand frustration and mistrust among the Unitron employees but on the other hand the realization of initial market synergies. The employees complained about the way of 'phonakizing' them and the lack of training, support and communication provided. This integration style and partial lack of leadership due to the absence of the CEO resulted in Unitron's loss of market share. Besides these problems the launch of the first fully digital product Nexus, which was supposed to come up for the acquisition investment, was delayed due to technical problems on the DSP factory Toccata platform on the supplier side and other manufacturing difficulties. During this delay of about six months, competition flooded the market with similar products. Thus the Nexus product – introduced too late and still with technical problems – had never really become a success.

At the same time Phonak in Stäfa noticed that the market for the medium-range products had shifted drastically. Digital technology was being launched even in lower-range products by the competitors, thus a product for the medium segment with full digital technology was due within 18 months. This trend had been overlooked mainly due to the high involvement of top management in integration issues. One manager described the situation thus: 'we had too many cooks and nobody served the food'. Thus the idea came to use the supposedly cheaper DSP factory signal-processing platform Toccata and Unitron's open-platform programming skills and knowledge to develop a new Phonak product for the mid-range market – the Aero. The product concepts can be seen in Figure 4.19.

Due to the initial difficulties at Unitron, the employees had the feeling that Phonak was taking their technology away and competitively intruding on their market. However, due to a highly capable project manager who came from Phonak but spent much of his time at Unitron, the resentments

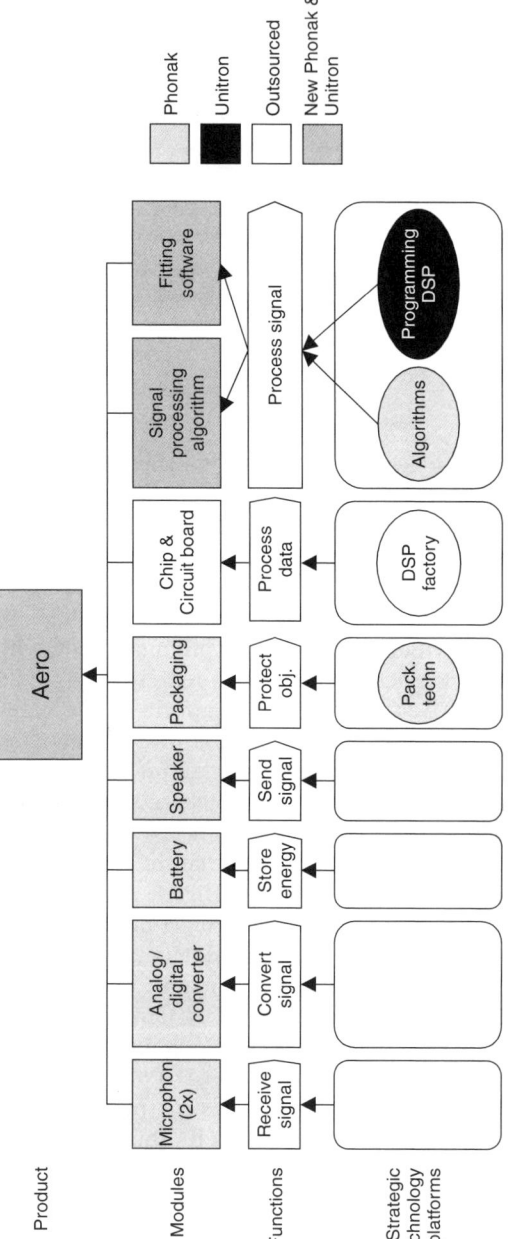

Figure 4.19 Product concept of the Aero

were reduced and cooperation was enabled; however, it remained difficult to achieve due to a lack of boundary-spanning structures. Surprisingly this project manager had not even had a great deal of technical expertise but he was reported to be very good at managing and guiding people. The Aero product was a Phonak product based on the DSP factory Toccata platform and some software elements from former Phonak products. Software was written in joint cooperation between Phonak and Unitron as only Unitron had the competence to programme the DSP factory Toccata. However, despite some improvements on Toccata, the platform was still having some technical problems and functionality was partly suffering, so that the Aero product initially was not a success. Only after some revisions did the product become highly successful for Phonak. The Aero was launched in October 2001, the result of a very fast product development process that was, possible only due to the shared new technology base of Phonak and Unitron.

After these first integration experiences, Unitron's loss in market share, the pitfalls of the Nexus product and some difficulties with the development of the Aero product, the awareness rose that this at-arm's-length integration approach might not be as appropriate as initially thought.

Thus the second integration phase – *the restructuring phase* – emerged at the end of 2001. It was initiated by the next *product development project* for Unitron's mid-range market – the Unison. This product was initially planned for production by both companies; however, due to a large workload in Stäfa, Unison turned out as a Unitron product. The project was launched by Phonak USA's CEO in line with the *new integration strategy*: '*one kitchen – two restaurants*'. This integration, and also business strategy, emphasized the focus on shared use of competencies and fostering innovativeness. The Unison product should contain a new externally sourced signal-processing platform due to the difficulties with the DSP factory Toccata platform. Furthermore, the Unitron algorithms and the packaging which were already developed and in use at Phonak should become part of the Unison product (see product configuration Figure 4.20). This leveraging of the packaging implied that not only product technologies but also process technologies such as CAD systems needed to be leveraged. The new awareness of tighter cooperation resulted in shifting three top engineers from Stäfa to Unitron to support the interface between designing and constructing Unison and adapting and manufacturing the packaging in Stäfa. The Unison product was launched in June 2002 and became a great success for Unitron.

During the product development project another drastic change marked the shift to this restructuring phase. The former CEO of Phonak USA became CEO of Unitron in Spring 2002, replacing the CEO of Phonak Holding who was seriously engaged in Switzerland, at this position, Phonak employees regretted this move but management admitted: 'At that time it was either sacrifice of the head of Phonak USA or losing the whole Unitron Company.' This new CEO of Unitron, as an American and expert in the hearing-aid

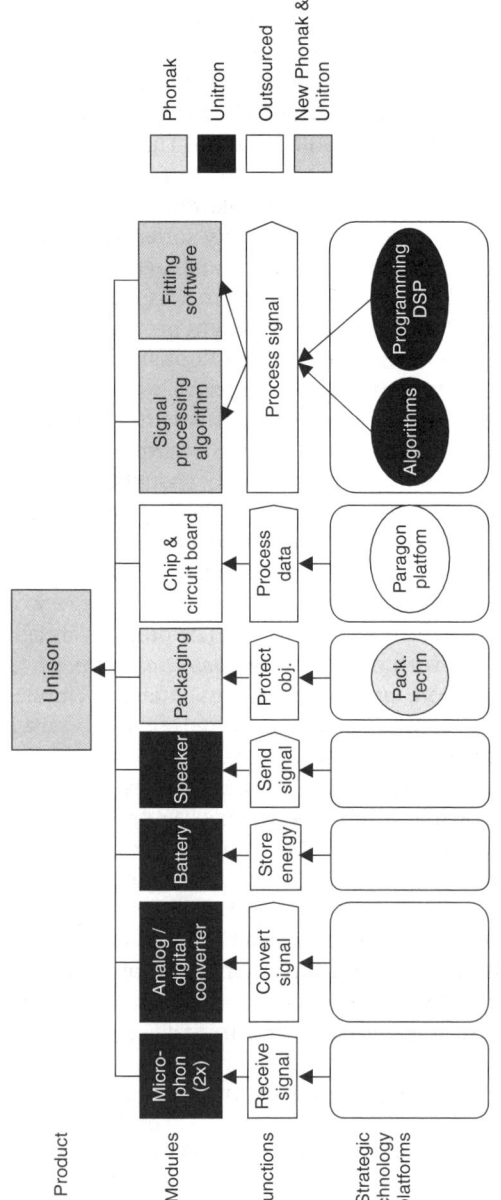

Figure 4.20 Product configuration of the Unison

business, managed the turnaround of Unitron and filled the leadership gap from which it had suffered the previous year. Another indication that the integration was attaining more attention was the official press release noting the previous difficulties: Phonak released the required one-time restructuring costs of CHF 12 million related to the Unitron integration and a revision of the overall EBITA which accounted for only 70 per cent of the original forecast. However, from then onwards the integration became tighter and was mastered in a Phonak-mindset.

Another change in Phonak, the assignment of a new CEO of Phonak Holding in October 2002, replacing the former CEO, brought another push to the integration of Unitron and thus initiated the third tighter integration phase. The new CEO of Phonak Holding publicly addressed the existing challenges faced by Phonak and Unitron together and he urged for an even tighter organizational integration, for example the establishment of shared services and a systematic approach for identifying and reaping technological synergies. Also future products, such as the Perseo and Conversa, integrate technologies from the merged resource base and can thus best serve specific market needs. To finally overcome the integration issues another write-off of CHF75 million, equal to about one-half of Unitron goodwill, was announced. This open step and the upcoming gains in market share due to the Unison product were rewarded by the stock exchange after announcement of a jump by 15 per cent.[11]

4.3.3.5 Interrelations, conclusions and discussions

Now the question arises of what can be learned from this case to improve the understanding of technology-based value creation in corporate acquisitions. The discussion will be guided by the relevant aspects identified within the state of the art in theory.

Generally the *Unitron acquisition* can be described as a friendly, relatively large acquisition of a closely related company, a direct competitor with a complementary market share and partly similar and also supplementary technologies. The acquisition took place in a highly dynamic marketplace, and the technologies finally leveraged and exchanged can be characterized as key technologies. Interestingly technology-based value creation was not pursued initially in the acquisition but emerged as an important and partly necessary source of value creation. Technologies were mainly leveraged to launch new innovative products. Resource deployment, however, happened only at the transfer of the manufacturing of Unitron's packaging to Phonak. More in-depth transfer and efficient deployment of technologies was initiated only in the beginning of the third integration phase by a tighter organizational integration pushed by the new CEO in 2003.

The *initial conditions* of the acquisition can be described as quite favourable and conducive to technology-based value creation. Unitron was a very good general strategic match as well as a good technology-strategic match. It gave

Phonak the opportunities to internalize knowledge on programming a fully open digital signal-processing platform in addition to its knowledge with its proprietary systems. Furthermore, the two companies had quite similar organizational, cultural and dominant business logic contexts. Unitron was described as more agile and faster in decision-making; an aspect which was honoured and respected by Phonak management. Thus contextual aspects did not hinder the integration progress and technology-based value creation. Only the lack of acquisition experience and of a formalized M&A support resulted in a non-integrative and non-holistic acquisition and integration process. Furthermore, Phonak at that time did not feel the urge to drastically change itself in the course of the business, an aspect which now in 2003, while Phonak is in the course of strategic and cultural change, is regretted. It was said that Phonak, while giving Unitron a new start to survive in its future, missed the opportunity to change accordingly. Another lack of initial conditions was the less-than-attractive technology potential of Unitron, especially due to the problems with the spin-off of DSP factory and bought-in Toccata platform. Thus the technologies were only partly attractive and not sufficiently mastered.

The *acquisition and integration processes* did not follow a standardized procedure and thus were marked by some shortcomings. One main difficulty was that the processes were not managed in a very integrative, systemic and holistic way. Thus only a few people knew about the strategies, and were involved only sporadically. Furthermore, integration planning was barely addressed during the acquisition phase. The technology due diligence was well mastered and even hinted at the potential technical problems but, however, these were paid insufficient attention later on, probably because the technology aspects were barely integrated into the overall acquisition decision-making process. The integration process, as described with three phases, made it obvious that technology-based value creation requires a strong leadership, some extending of organizational boundaries, trust-building phases and job rotations or people transfer. Furthermore, the restructuring phase managed by the CEO of Phonak USA applied a very direct approach that was not supported by change-management measures such as extensive communication, training and the like. These tough, however partly necessary, actions created mistrust within the Unitron company which had a direct effect on the technology-based value creation.

The *external developments* also had an important impact on technology-based value creation. Due to Phonak's focus on the integration task, the external developments were overlooked and the rapid shift in market was recognized almost too late. Thus on the one hand these external developments pushed Phonak behind the competitors and had a very negative overall effect, while on the other hand the subsequent urge to launch a new product in a very short time fostered joint product development projects, cooperation and finally trust building. Thus in this case the external developments gave rise to reaping synergies, therefore it seems that pressure from outside helped to meld the companies together.

This case is a good example which shows the interrelation between acquisition type, initial conditions, external developments and strategy processes. Only if all play in accordance and support each other can technology-based value creation be achieved.

4.3.4 Case 4: Unaxis–Plasma-Therm acquisition

The Oerlikon Bührle Group (OBH) announced in December 1999 its intention to acquire Plasma-Therm, Inc., a manufacturer of semiconductor equipment in St Petersburg, Florida, US. After receiving clearance from the regulatory authorities, the deal closed on 7th of February 2000. Oerlikon Bührle acquired 95 per cent of Plasma-Therm's 11,252,311 issued shares at a price of $12.50 per share, which amounted to approximately USD 150 m.

The acquisition was in line with OBH's strategy of strengthening its position in the semiconductor equipment business. The companies complemented each other regarding market access and technological competencies. However, due to integration problems and a rapid market growth the technology-based value creation objectives were quite delayed. In general the acquisition created high-value creation potential but, however, this potential could not be fully exploited for several years (for the acquisition timeline see Figure 4.21).

4.3.4.1 Initial situation

OBH Holding, renamed Unaxis Group in early 2000, has a multifaceted history. Founded in 1906 it pursued a diversification strategy resulting in a company employing 37,000 people in 1980 and manufacturing a variety of products ranging from aeroplanes, optical elements, shoes and weapons to mortgage services. A shift in strategy towards a stronger focus on thin-film technologies initiated several divestitures until 1999 and is partly still ongoing. Additionally, Unaxis strengthened its competencies via acquisitions in the high-tech sector, such as the acquisition of the Leybold Company specializing in vacuum technologies.

In 1999 Unaxis consisted of three main segments (see Figure 4.22). Furthermore, it held an equity stake of 26.9 per cent in ESEC and owned additional small companies. The focus of this case lies on the information technology segment and in particular on the semiconductor business unit.

The semiconductor unit within the IT division already had a quite long history. It became independent in 1992. In 1999 the unit had revenue of CHF 83 m, an increase of 12 per cent from the previous year. The unit developed, designed, manufactured and sold equipment for the semiconductor industry. It mainly focused on certain niche markets within this highly competitive field: these were niches in the semiconductor front-end market, the telecom and sensor market, often also referred to as compound market,

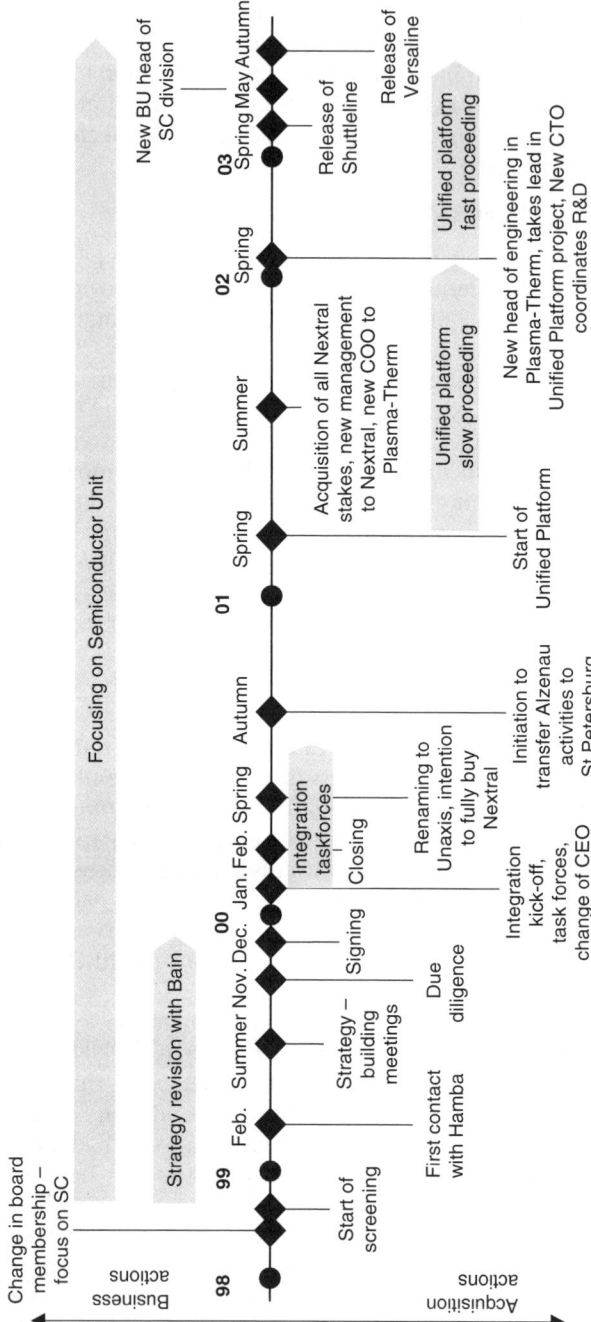

Figure 4.21 Acquisition timeline of Unaxis's Plasma-Therm acquisition

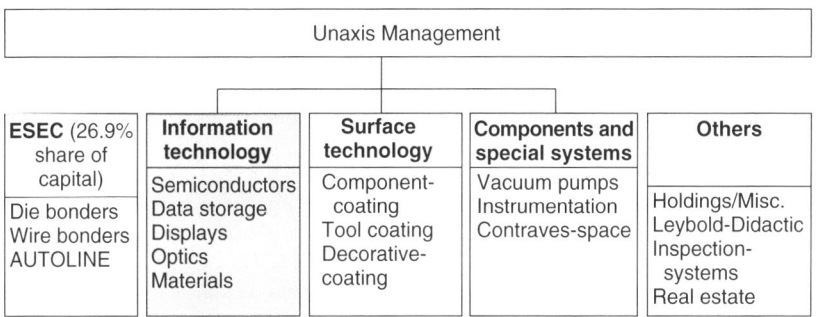

Figure 4.22 Unaxis's organizational structure in 1999

and the packaging market. All products sold within these markets were concerned with the deposition and removal of materials on wafers. They were engineered and manufactured at two different sites in Switzerland and France respectively. In Trübbach, where the headquarters of the unit was also located, cluster machines called CLC 200 and CLC 300 were produced. These high-end and large-scale machines could be used for various applications for all three markets. Furthermore, batch machines[12] such as the LLS EVO, the BAK and the Sirius machines, were produced for other specific applications. The main competence of the engineering division in Trübbach lay in the sputtering technology also referred to as PVD (physical vapour deposition) (see also the innovation architecture in Figure 4.23). Furthermore, the engineers in Trübbach mastered the CVD (chemical vapour deposition) technology for applying SiGe, a material mainly used in the semiconductor front-end market. The second engineering and manufacturing site was located in Grenoble, France. It was an acquisition of a stake of the Nextral company. Nextral also manufactured equipment for the semiconductor industry, but was focused on the compound segment. These machines, called NE and ND, were much smaller than the CLC machine and followed a batch process. The machines were based on the PECVD (plasma-enhanced chemical vapour deposition) technology. Furthermore, these systems were based on the Plasmabox a highly specialized process module patented by Unaxis. In addition, Nextral developed the soon-to-be-launched Quadra system, a small cluster machine.

The semiconductor (SC) unit was structured according to markets in front-end products, telecom and sensors, and packaging. Each of these units consisted of its own sales, manufacturing and engineering function, following the market orientation emphasized by the CEO of the Unaxis Holding.

Besides the semiconductor unit another part of the Unaxis company played an important part in this case study. The Leybold company, acquired by Unaxis and partly located in Alzenau, Germany, developed and manufactured

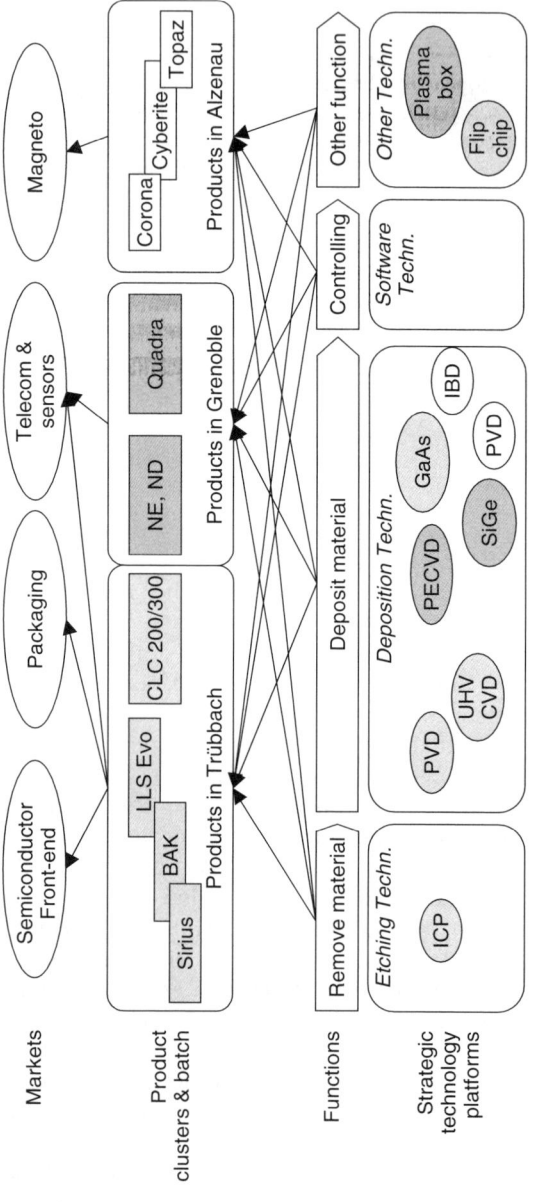

Figure 4.23 Rough innovation architecture of the SC unit and Leybold's magneto business

equipment for the magneto market (products such as Corona, Topaz, Cyberite), and the hard disc market (products such as Circulus M Series, Emerald Series). These products for the magneto market were based on PVD and IBD (ion beam deposition) technology.

In 1998 one board member of the Unaxis Holding changed and initiated a stronger focus on the semiconductor business unit. As a consequence, the consultancy Bain & Company was hired to support the process of finding growth opportunities to strengthen this business. As a first step Bain & Company outlined the upcoming trends, such as an increasing demand for specific chips for digital camera or mobile phone applications, new technologies and a strong industry consolidation. Furthermore, they developed four growth options, such as:

(1) Maximizing the existing core business;
(2) Adding a neighbour niche within the thin-film technologies;
(3) Entering the mainstream market within the thin-film business;
(4) Entering non-thin-film related applications.

The four options were analysed and diligently compared. The concluding recommendation was to stick to the quite profitable niche markets within the semiconductor industry and to add other, however related, thin-film technologies and competencies. Figure 4.24 provides an overview of the thin-film-related areas within the niche market of the Unaxis semiconductor business. Thus the semiconductor management was assigned the task of screening for potential acquisition candidates. Unaxis at that time did not have a corporate M&A team or an established acquisition process.

Very soon the Plasma-Therm company was identified as a potential acquisition target by the SC's CEO. This company was not new to Unaxis. On the one hand it had always been viewed as a potential strategic acquisition target, although there was no focus to grow in the semiconductor segment at that time, and on the other hand one business unit manager of Leybold was a member of the board of Plasma-Therm.

4.3.4.2 The Plasma-Therm Company

Plasma-Therm, founded in 1975, was also engaged in the design and production of thin-film etching and deposition manufacturing equipment, although for partly different applications and markets compared to those of Unaxis. The company's headquarters was located in St Petersburg, Florida, US. In 1999 the sales volume reached CHF 61 m, a decrease of 17 per cent from the previous year, caused by the worldwide downturn in the semiconductor equipment industry in 1999. The company employed 186 employees of whom 148 were located in Florida and of these 42 were engaged in the R&D unit. Plasma-Therm was serving four different markets: optoelectronics and telecommunications, MEMS (micro-electronic mechanical systems),

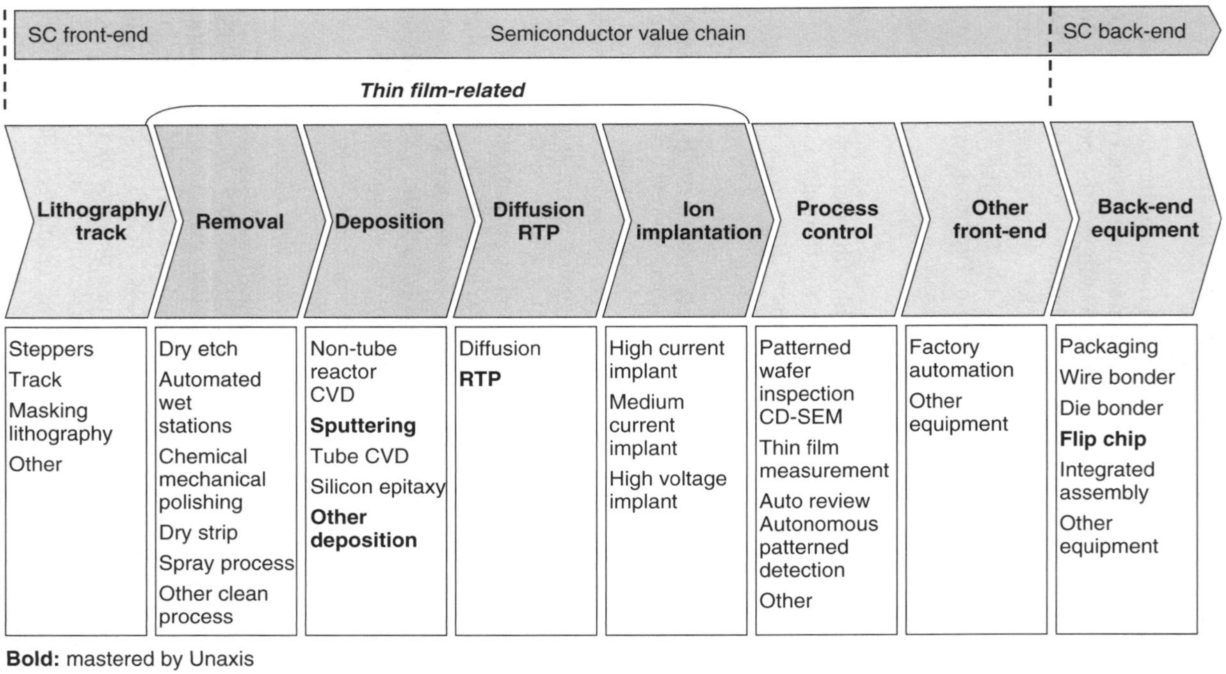

74

SC front-end	Semiconductor value chain							SC back-end
	Thin film-related							
	Lithography/ track	**Removal**	**Deposition**	**Diffusion RTP**	**Ion implantation**	**Process control**	**Other front-end**	**Back-end equipment**
	Steppers Track Masking lithography Other	Dry etch Automated wet stations Chemical mechanical polishing Dry strip Spray process Other clean process	Non-tube reactor CVD **Sputtering** Tube CVD Silicon epitaxy **Other deposition**	Diffusion **RTP**	High current implant Medium current implant High voltage implant	Patterned wafer inspection CD-SEM Thin film measurement Auto review Autonomous patterned detection Other	Factory automation Other equipment	Packaging Wire bonder Die bonder **Flip chip** Integrated assembly Other equipment

Bold: mastered by Unaxis

Figure 4.24 Thin film-related competencies within the semiconductor value chain

data storage and especially photomasking. Plasma-Therm served these markets with mainly three machines: the small 700 Series, which could be used for small R&D applications only; the Shuttlelock machine, also tailored towards R&D applications; and the medium-sized Versalock, designed to serve the needs of industrial applications within the four markets. The core technology of Plasma-Therm was PECVD technology, which was used to deposit non-silicon materials on to wafers for various applications. Additionally, it excelled through various other complementing technologies such as etching technologies (RIE – reactive ion etching, ICP – inductively coupled plasma,) or HDP (high-density plasma).

Plasma-Therm had partly very similar customers to those of Unaxis, especially within the compound areas; however, Plasma-Therm's focus was more within the US. Furthermore, the technologies were highly complementary as Plasma-Therm provided machines to etch the masking for lithography while Unaxis was more focused on subsequent steps within the value-chain (see Figure 4.24). Additionally, Plasma-Therm's technologies were highly attractive and new, for example the CVD technologies were to partly replace the sputtering technologies.

Due to the crisis within the semiconductor industry, the owners of Plasma-Therm wanted to sell the company. This seemed to be a perfect match between Unaxis's and Plasma-Therm's intentions.

4.3.4.3 The transaction phase

Generally the transaction phase can be divided into two phases: whereas the first lasted from February to October 1999 and was concerned with strategic and technological aspects, the second phase lasted only during November and December and was concerned with legal, financial and tax issues.

On 17 February the CEO of the SC division first visited Plasma-Therm in Florida. When he returned he set up an acquisition team to pursue the acquisition of Plasma-Therm. The team consisted of the SC CEO, a project coordinator, an M&A expert and consultants from Bain & Company. Additionally, a steering committee was set up to accompany the acquisition, consisting of the Unaxis Holding CEO, the SC CEO and some people from the controlling department. The objective of the acquisition was to gain access to the technological competencies of Plasma-Therm and particularly to the engineers. Furthermore, the acquisition could help the semiconductor division increase its market share in the US and achieve a critical size to gain increasing power and work with scale effects.[13] The acquisition process was well guided by Bain & Company (see Figure 4.25).

During the summer of 1999 the acquisition team of Unaxis visited Plasma-Therm twice. The Plasma-Therm company was introduced and visited, the strategy was challenged, the key people were identified and the technologies were investigated. As a result of this more strategic due

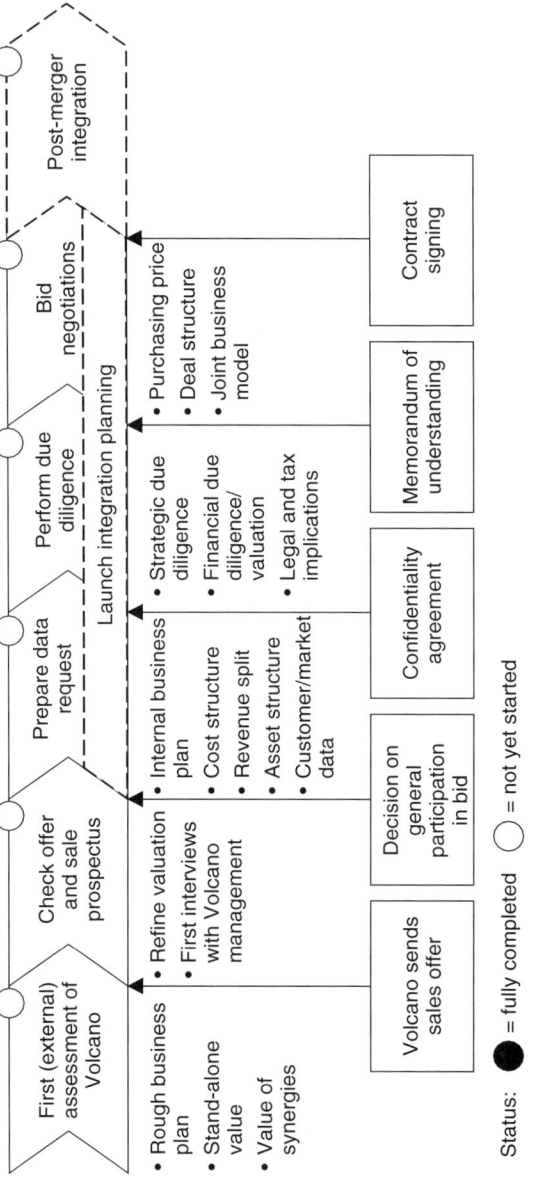

Figure 4.25 Acquisition process of the Plasma-Therm acquisition

diligence, Unaxis found out that:

- Plasma-Therm intended to enter the Unaxis PVD-technologies;
- Several technology development projects were ongoing;
- Plasma-Therm intended to build a new building;
- The cultures of Plasma-Therm and Unaxis matched very well.

The investigations were concluded with the financial business model, which evaluated the value-creation opportunities (see Figure 4.26).

During November and December the lawyers, financial experts and others conducted their due diligence. During this time Bain also made some proposals for the post-merger acquisition phase – the integration and its planning. They recommended consolidating the technologies into centres of competence. Furthermore, they recommended a quite fast and direct integration approach. This was in line with the intentions of the semiconductor unit, especially as Unaxis had negative experiences from not integrating acquired companies. The CEO of Unaxis Holding said accordingly: 'If we cannot fully integrate a company, then we don't buy it.'

However, when Bain left the project, by the end of November, the integration planning was neither detailed nor fully outlined.

4.3.4.4 The integration process

The integration was initiated by the kick-off meeting on 4 January 2000. The SC CEO established eleven integration teams, one core team and also a steering committee. All teams comprised people of both companies equally. These teams were assigned the tasks of establishing joint sales forces, restructuring the organization, retaining the key people, and so on (see Figure 4.27). The headquarter of the Unaxis semiconductor unit was transferred to St Petersburg, Florida and the SC CEO replaced the former CEO of Plasma-Therm, who had to leave for personal reasons. Furthermore, the organizational structure of Plasma-Therm, which had only one centralized R&D unit, was adapted towards the Unaxis organizational design which was characterized by a very strong customer focus and thus a decentralized R&D organization. The integration was done in a very cooperative and gentle way. This was justified by the perfect match between Unaxis and Plasma-Therm and the willingness of the Plasma-Therm employees to be bought by Unaxis. However, this initially very promising approach (most of the integration teams were finished within eight weeks) also had its disadvantages, especially with the integration of the two different SAP systems. The negotiations and discussions hindered the SAP integration, and enthusiasm about the acquisition soon turned into frustration. Furthermore, the soft and distant integration of the business processes resulted in unclear interface between the operations and within the organizational structure itself, which also hindered daily business, interaction and communication. Within this phase, the

Revenue synergies	Cost synergies	Soft synergies	Financial synergies		Total CHF m
• Cross-selling TBB PVD products to STP USA & photo mask customers • Sell both products to Asia • Cross-selling STP Etching to TBB customers • Sell service contracts to STP customers	• Save demo lab for TBB • Save sales provision • Consolidate sales & service in US • Consolidate STP and Nextral products • Merge Versalock and Quadra platform • Transfer Volcano PVD to TBB • Consolidate overhead functions • Consolidate suppliers	• Grand leverage • Strengthened market presence and increased sales penetration • Know-how sharing along value chain steps • High cultural fit	• Tax leverage possible of STP earnings in TBB structure		
Revenue impact 03 28	–	Not quantified	–		28
EBIT impact 03 2.9	7.5	Not quantified	To be defined		10.4

Figure 4.26 Value-creation opportunities in the Plasma-Therm acquisition

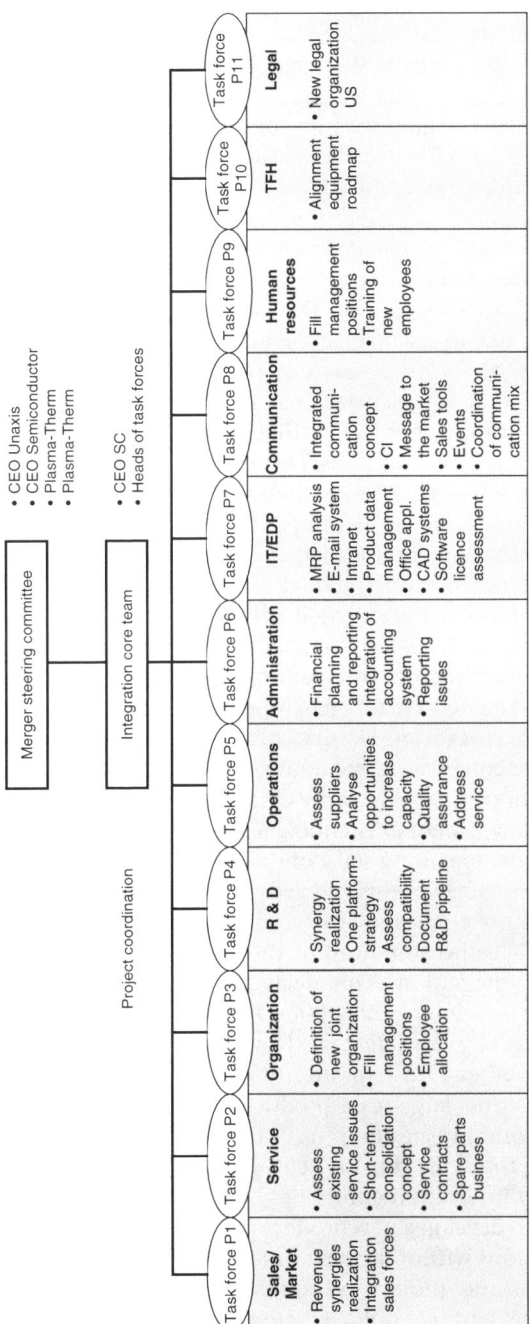

Figure 4.27 Integration teams of the Plasma-Therm acquisition

economy started to take off again, thus all resources were engaged with satisfying immediate customer needs. Thus technology integration could not take place within this first integration phase of approximately eight weeks.

After this initial integration and consolidation effort, the technology integration phase started. However, as the SC CEO was bound between two locations in Florida and Trübbach and the immense workload persisted, technology integration became quite difficult. Furthermore, the strict market-oriented organizational structure within the unit hindered technology transfer and cooperation.

In a first step towards transfer the complementary, joint assets of the Leybold part, located in Alzenau, and Plasma-Therm had to be reduced. This project came up about three-quarters of a year after the acquisition in autumn of 2000.[14] It was decided that the magneto business in Alzenau, which was not running very well at that time, had to be moved to Florida. This transfer was done in two steps. First of all the machines and the people were transferred, whereas the most important engineers could be retained and then transferred later. Furthermore, the head of the magneto business in Alzenau became the head of this merged business at Plasma-Therm. The second step in this transfer was product and technology integration. In the beginning, the products were aligned and afterwards a completely new and joint product line was initiated based on the merged technology platforms.

The subsequent technology integration project aimed to reduce the redundancies between the new Quadra machine from Grenoble and the Versalock and also the 700 Series from Plasma-Therm by merging them on one joint platform. The machines were medium-sized and aimed for the compound market, Furthermore, they were partly based on the same technology: PECVD. However, the management at Grenoble resisted this effort. Therefore Unaxis decided to buy the remaining stake of the Nextral company in order to gain full control and achieve the merged platform. This legal transaction, however, took more than one year – until summer 2000. Finally, another ex-Unaxis employee became head of the Nextral company and facilitated the technology integration. This delay was the main factor among the reasons why the technology integration between Plasma-Therm and Unaxis was achieved only at a very late stage. However, when the Nextral was finally fully bought, another idea had already come up. The idea was to merge several medium- and large-scale products on the same unified platform. Whereas the second integration project to merge the Quadra machine with Versalock never took off, this unified platform project became the main, third technology integration effort.

The first idea to develop a new product platform for several machines from the diverse locations within the new Unaxis semiconductor unit came up in mid-2000, when the project coordinator tried to match the different markets, products and technologies. However, due to the legal issues with

Grenoble, the development project was postponed until spring 2001. The objective of the unified platform project was mainly to reduce cost by basing several machines on the same modules and thus increase standardization, lower maintenance and purchasing costs, and reduce manufacturing expenses. Furthermore, a unified interface to the customer in all machines would increase the proposed value. Thus the Unified Platform should be a combination and consolidation of the Clusterline so far produced in Trübbach, the Versalock, the main machine of Plasma-Therm, the Cyberite machine from Alzenau and the Quadra machine of Nextral.

The project was headed by the Plasma-Therm head of manufacturing. He distinguished between software and hardware teams and assigned tasks on the product module level (see Figure 4.28). One of the main aspects was the definition of the core element of the Unified Platform, the handler in the middle of the system and the software which steers this central part and controls the decentralized modules.

All machines that had to be integrated were so far based on different handlers; whereas Plasma-Therm had one in-house-made handler, Unaxis based the CLC machines on the Brooks platform bought externally. Very soon it was decided that the main handler would be the Brooks system and that common software would be used.[15] So far the machines had been based on quite similar software; however, different versions were applied which made the direct integration quite complicated. The software teams from Trübbach and St Petersburg could work together quite well, especially as they knew each other already from a conference in Hawaii. The hardware teams had to struggle with the interfaces between the handler and the different modules. All engaged units and people demanded different requirements from the system and thus conflicts of interest appeared. As the project manager did not have any boundary-spanning power to easily consolidate the different opinions, the project was often led in an unassertive way and thus did not proceed very rapidly.

Besides the slow technological integration, the economic upturn experienced during the first phases of the integration turned into an extensive downturn and Unaxis was forced to lay off some of the workforce. Even though Plasma-Therm was bought mainly because of its excellent competencies, it was mostly people in St. Petersburg who were dismissed, whereas in Trübbach the workforce remained practically intact. Furthermore, an employee survey in St Petersburg showed that motivation and enthusiasm were very low, probably also the result of the continuous absence of direct leadership. These alarming developments attracted the attention of the Unaxis management and initiated several changes within the semiconductor division from the summer of 2001 onwards. Plasma-Therm was assigned a new COO.[16] Furthermore, the engineering divisions within the semiconductor units became more centralized, which immediately facilitated technology integration. Additionally, at the beginning of 2002, Plasma-Therm

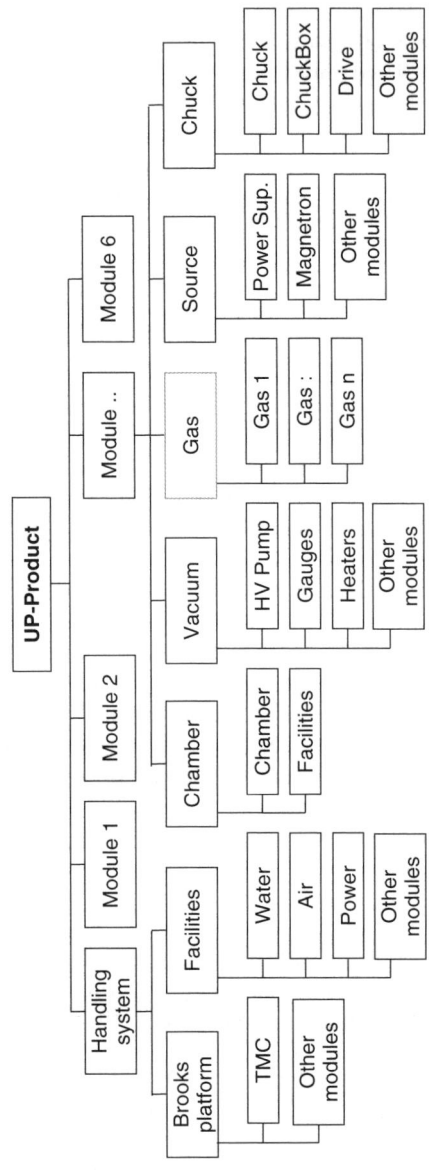

Figure 4.28 Modular structure of a Unified Platform product

got a new head of engineering, who took over the Unified Platform project, and additionally a CTO, who was in charge of facilitating boundary-spanning cooperation and technology integration. As the new head of engineering was also part of the management team, the different requirements and interests of the various units and locations could be managed more directly and easily. From then on the technology integration and thus the Unified Platform project became very high priority and proceeded more quickly.

A detailed roadmap of the projects was developed and the effective deployment of the technology groups was achieved (see Figure 4.29). Finally, the various technological redundancies could be abolished and centres of competence could be established. Whereas the engineering and manufacturing in Trübbach now excels through its knowledge within the semiconductor front-end market and accordingly masters the PVD and CVD for SiGe technologies, the engineers in Plasma-Therm have now centralized all knowledge within the non-silicon, compound, magneto and photomasking areas. Alzenau was completely transferred to St Petersburg and Grenoble became like a satellite to Trübbach. After this redeployment of capabilities and the platform development project, the product development projects took off and the innovation rate rose. For example, the new Shuttleline was launched in March 2003 and the new Versaline followed in Autumn 2003. These successes would not have been possible without the deployment of the joint technology bases of Plasma-Therm and Unaxis Semiconductor. Furthermore, joint innovation processes and more profound IP management were established. The integration effort will be finalized by the new head of the semiconductor division whose position became effective in May 2003. Additionally, a corporate M&A team was established which would support the acquisition process. This team, consisting of five persons in 2002, is supported by the corporate CTO especially to validate the soundness of the technological competencies and patents of the target. This corporate M&A team is so far not involved in the integration phase and entirely focuses on the appropriate proceeding within the acquisition process.

4.3.4.5 Interrelations, conclusions and discussions

What interrelations between the technology integration and other aspects within the acquisition and integration can be found?

Generally the *acquisition* can be described as a friendly, fairly large acquisition which allowed the internalization of complementary, supplementary and partly redundant technologies and the access to a larger geographic scope of the existing customer base. The acquisition took place in highly turbulent times within the semiconductor and semiconductor equipment industry. The acquired competencies were pace-maker and key technologies and thus increased the technology potential of Unaxis.

Most of the *initial conditions* before the acquisition pointed to a very promising result. The lack of a formalized M&A team and much acquisition experience

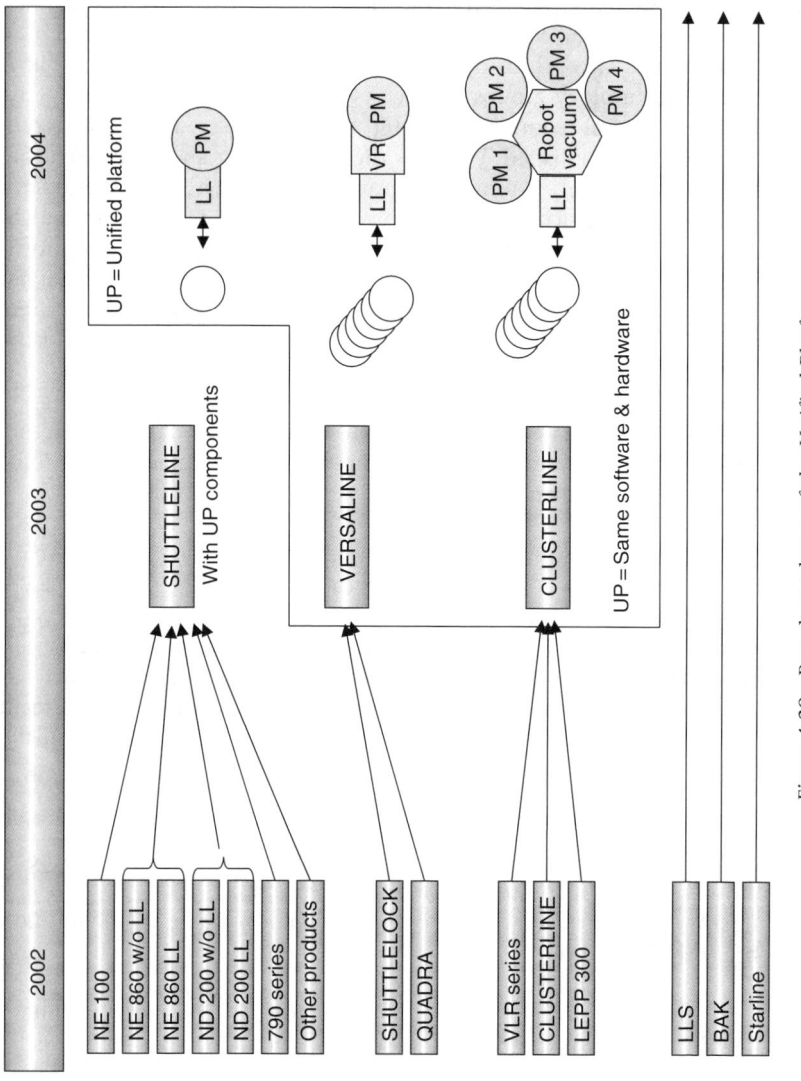

Figure 4.29 Rough roadmap of the Unified Platform

was abrogated by the presence of the consultancy Bain & Company which led the acquisition process very well. Furthermore, Unaxis was highly familiar with the technologies and partly even with the employees of Plasma-Therm and thus it was equipped with sufficient absorptive capacity to assess and integrate the acquired competencies. Also, from the strategic perspective, Plasma-Therm fitted very well with the growth, business unit and technology strategy. The PECVD technology was an especially attractive buy-in. Thus the value-creation potential to reap resource deployment and also innovations was quite large. The only hindering characteristics of the two companies were that there seemed to be a perfect contextual match between the two companies which reduced the sense of urgency during the integration phase. Thus a very good match in organizational characteristics, company culture and dominant business logic falsely implied that integration would proceed very easy. Furthermore, even though the semiconductor unit showed willingness to change itself by transferring the business unit headquarters to St Petersburg, the leading managers were permanently moving between Florida and Switzerland. This lack of direct leadership influenced the whole integration process and thus also the technology-based value creation.

The main impact on technology-based value creation within the Plasma-Therm acquisition rooted from the process design of the *acquisition and integration process*. The acquisition process was well mastered due to the support of Bain & Company. Within the due diligence only were some IP-related issues not identified which finally caused the discontinuation of one product; however, generally no technology-related surprises occurred after the acquisition. The integration process was not so well conducted. The cooperative and soft integration behaviour especially in the beginning, when changing the company structure and the supporting systems, resulted in time delays and confusion. Furthermore, the lack of direct leadership during the technology integration phase caused long negotiations and a lack of value creation. The establishment of boundary-spanning structures, such as a more centralized R&D organization and one new head of engineering as well as a new coordinating CTO, facilitated technology transfer and cooperation and thus technology-based value creation.

The *external developments* played a dual role within the acquisition. In the first integration period the drastic increase in demand helped to make the acquisition a financial success quite soon after closing; however, technology integration could not be pursued due to a lack of resources. In the subsequent phase, when the economy abruptly slowed down, time was provided to redeploy the technologies and develop the new product platform.

Therefore again this case shows the close interrelation between acquisition type, initial conditions, external developments and the strategy processes. It can be inferred that technology-based value creation requires clear direction and leadership, direct strategic attention, boundary-spanning structures and the appropriate external pressure but also some slack. Furthermore, a soft

4.3.5 Case 5: Starrag–Heckert acquisition

The manufacturer of high-tech five-axe milling machines, Starrag, located in Rorschacherberg at Lake Constance, Switzerland, acquired the German-based Heckert on 1 August 1998. Heckert was engaged in the engineering, manufacturing and sales of four-axe milling machines and mainly supplied the automotive and land-machine markets. The acquisition offered both companies the opportunity to reposition themselves within the changing industrial environment. In the course of the integration, the conjoined companies excelled in merging their technologies and in adopting a new product platform strategy which allowed them to reap extensive technology-based synergies in terms of reduction of production costs and an increase in innovativeness. In addition, the companies could achieve cost reductions through scale effects within purchasing and sales. The successful technology-based value creation was made possible by the profound strategic and technological fit of the two companies and the direct, supported, boundary-spanning integration approach (for the acquisition timeline see Figure 4.30).

4.3.5.1 Initial situation

Starrag was founded as the Henry Levy factory in 1897. At that time the company produced threading machines for stitchery businesses. After the severe crisis in the textile industry, the company started to focus on machine tools and soon focused on milling machines.

In 1998 Starrag employed around 400 people with revenue of approximately CHF 70 m of which 10–12 per cent was invested in R&D. Starrag was concerned with the engineering, manufacturing and sales of blade-milling machines, machining centres and flexible manufacturing systems for the aviation and aerospace industry, the generation plant and turbine industry and manufacturing businesses for tools and forms on a worldwide scale. It sold high-end five-axe machining centres (ZS, ZT Series) and blade-milling machines (NX, HX) to enable their customers to mill blades, blisks, turbines, impellers, and engine and structural parts. The machines were based on key technologies and were highly adaptable to customers' needs. Starrag excelled through offering a whole package as a complete solution for the customers, covering the machine, the CAM – software,[17] special milling tools and related technology. Starrag was organized according to functions and could be characterized by a quite conservative and traditional culture.

In 1998 Starrag was challenged by a change in industry needs (see Figure 4.31). Customers increasingly demanded smaller and cheaper machines. Furthermore, the existing product portfolio of Starrag came to the end of its life-cycle and the product pipeline was empty. Thus strategic decisions about which direction to pursue were due, however, not taken.

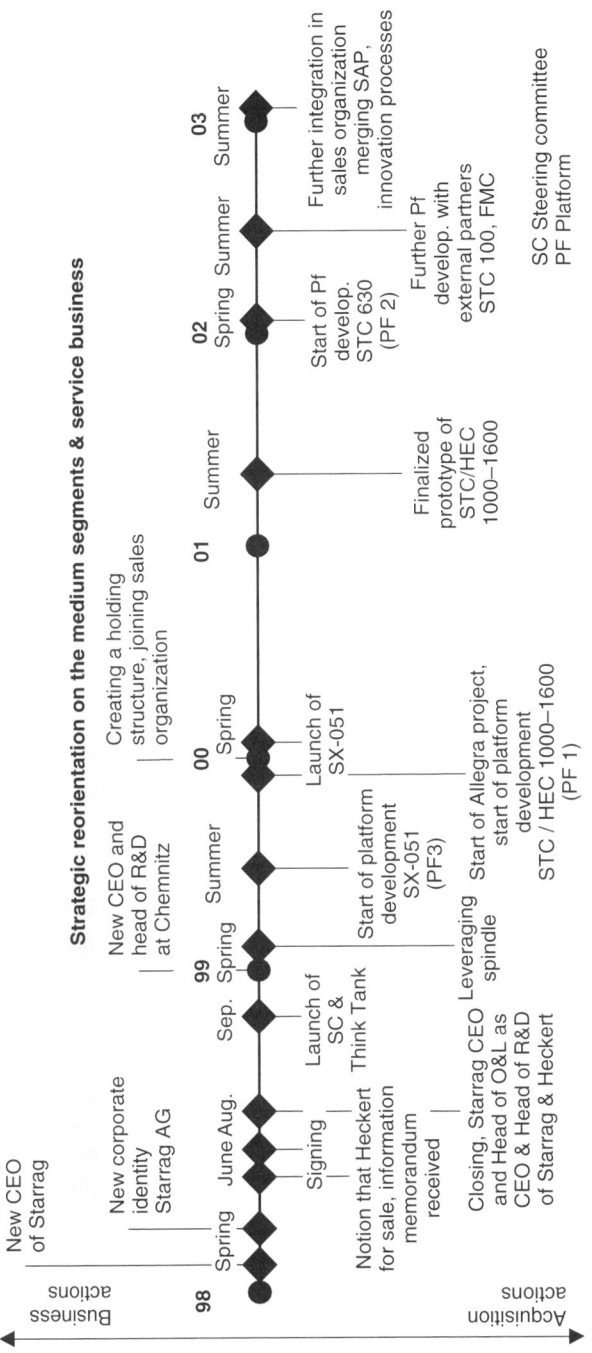

Figure 4.30 Acquisition timeline of Starrag's Heckert acquisition

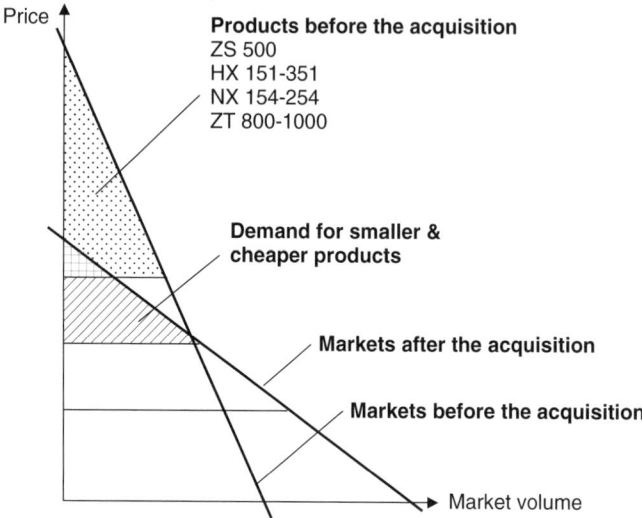

Figure 4.31 External developments within the Starrag markets

At the beginning of 1998 a new CEO was assigned to lead the Starrag AG. His task was to guide Starrag away from the upcoming crisis. It was clear to him, and he communicated this strongly, that Starrag had to access the medium segment of the industry and had to increase its focus on the service business; however, he knew that it would be difficult to transform a high-end 'Rolex'-type company into a performer within the middle-range sector: as a 'Swatch'-type company.

At the same time the Heckert company was in the middle of its receivership searching for potential buyers supported by the consultancy Roland Berger. As Starrag's CEO had good contacts within the industry and also with former East Germany, he knew Heckert quite well and proposed to his management team that they buy it in spring 1998.

4.3.5.2 The Heckert company

Heckert GmbH, headquartered in Chemnitz, Germany, has for a long time been an important player in the milling machine market. It was founded in 1885 as a bicycle producer, in 1934 it was already the largest milling machine producer in Europe, and in 1973 it manufactured 2,500 milling machines. In 1998 Heckert employed 380 people with revenue of approximately CHF 90 m and an R&D investment rate of 5–7 per cent. The company had been bought in 1993 by Traub but this acquisition was quite unsuccessful and resulted in a new founding of Heckert in 1997. In 1998 it had to go into receivership and searched for a potential buyer in spring 1998.

Heckert manufactured four-axe milling machines for the mainly German-based middle-range segments of the automotive, land-machine, production technology and engineering industry. With its products, the CWK line, it enabled the milling of gearboxes and parts as well as castings and other machine parts. Heckert excelled with highly competent engineers, cost-efficient engineering and manufacturing by keeping a very high-quality standard. Its machines were quite new and the pipeline was filled with new developments such as the SKM machine and the CWK 630D. Heckert was organized according to functions. Its culture was influenced by former East-European structures such as strong labour unions, medium sense of urgency, and so on. Due to its poor and disappointing last years, Heckert's employees were quite suspicious about upcoming changes; however, they knew that a change was the only way to keep the company running.

4.3.5.3 The acquisition process

Due to the short notice regarding Heckert's willingness to be bought in the beginning of June 1998, Starrag did not have much time to investigate the target or prepare the integration. The information memorandum was the main basis for the management team to briefly identify synergies and verify the strategic fit. The product lines complemented each other very well, as Heckert mainly covered the lower segment (see Figure 4.32). Furthermore, the acquisition supported Starrag in its business strategy to access the lower-range segment and to reap scale effects through growth. Additionally, the market segments did not really overlap and thus provided a good basis for achieving market and scale synergies. As Starrag's CEO was already quite

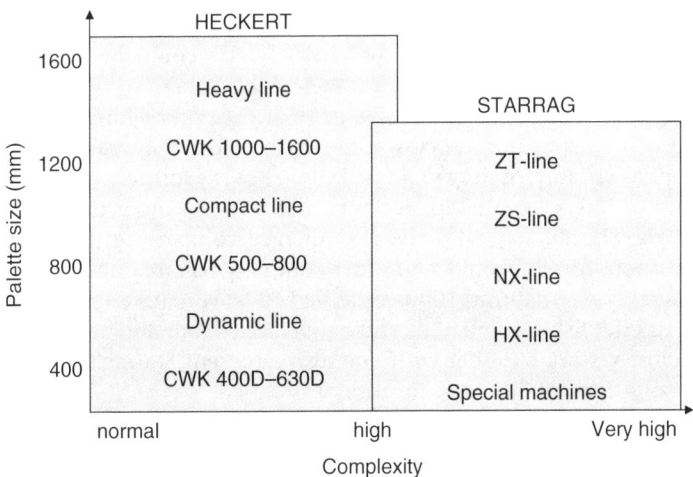

Figure 4.32 Fit of product portfolios between Starrag and Heckert

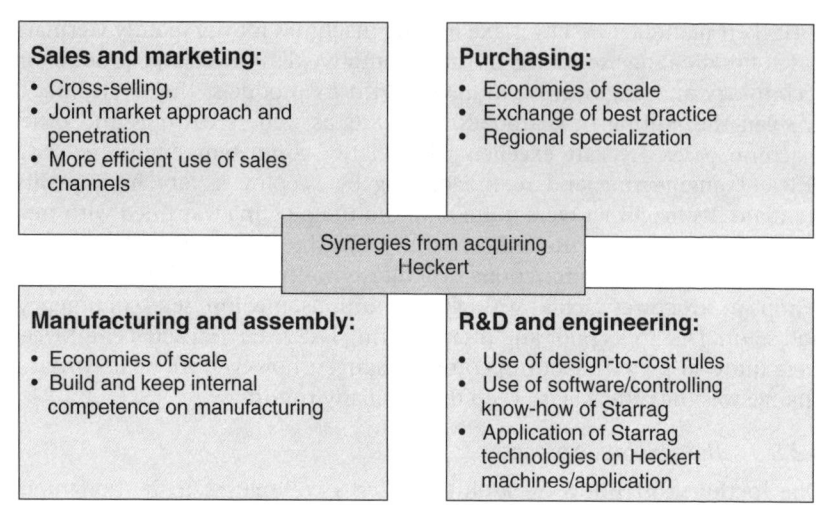

Figure 4.33 Expected synergies from acquiring Heckert

familiar with the Heckert company, products and competencies and as these were highly similar to Starrag's competencies, a technology due diligence did not seem necessary.

Starrag's management identified potential value-creation opportunities in four fields: sales and marketing, purchasing, R&D, and manufacturing and assembly (see Figure 4.33). Interestingly Starrag expected the technology-related knowledge transfer to occur mainly from Starrag to Heckert and not the other way a round. Thus Starrag was not that much aware of the huge technology-based value creation potential which could be derived from transferring Heckert's engineering know-how to Starrag.

The signing of the acquisition was done at the end of June 1998, which resulted in an acquisition process of merely two to three weeks. During the closing phase in July, Starrag's CEO asked one of his former colleagues from Germany to interview all 380 Heckert employees, in order to identify 190 people who should stay with the new merged company, and additionally organized jobs for 90 people at another company who bought a small stake in Heckert. The remaining 100 people had to leave. These interviews were aiming to establish a competent and cooperative team at Chemnitz, which would closely work together with and also represent Starrag's mindset. On the first of August 1998 the acquisition was closed.

4.3.5.4 The integration process

On the second of August, Starrag's CEO and Starrag's head of Operations and Logistics (O&L), who was responsible for not only the whole R&D but also manufacturing and other operations, travelled to Chemnitz and immediately took over operations. Whereas the head of O&L was initially assigned to support the CEO, it soon became clear that the head of R&D at Heckert needed to be replaced and thus the head of O&L took over both R&D sites. The objective was to bring Heckert back on track, to jointly approach the middle- and upper-market segments with an enlarged product portfolio, to increase the overall customer focus and to reduce the production depth at Starrag. The CEO and head of O&L pursued a very committed, strict and direct integration approach. For more than five months they spent a good deal of their time in Chemnitz, made quick decisions about change and supported the integration efforts by extensive communication. Furthermore, they invited the Heckert employees to visit Starrag at the Rorschacherberg. The individual teams introduced their work to each other and had the first opportunities to discuss potential synergies. Furthermore, the informal get-together in the evening also helped to break the ice between the two companies.

Besides the integration efforts at Chemnitz, the management and subsequently also several employees of Starrag held workshops on their own strategic reorientation and need to change towards a middle segment player partly away from their upper niche markets. These change efforts in Rorschacherberg and the direct integration leadership and efforts in Chemnitz together promoted the effective merging of the two companies.

From the organizational point of view, the CEO established steering committees and a Think Tank in September 1998 to support the integration efforts (see Figure 4.34). The steering committees were permanent, operational teams which embraced both companies. Depending on the existing requirements and competencies, these committees were headed by either Starrag or Heckert employees. They were in charge of the operational integration of areas such as IT and communications, sales and marketing, customer service, quality systems, purchasing and procurement, application engineering and innovation, and personal and organizational. During their integration projects they soon recognized that the expected synergies from economies of scale within the purchasing proved correct especially due to very good purchasing conditions with East-European countries and highly increased purchasing volume. Furthermore, they noticed that Heckert's sales channels were mainly focused on Germany, which provided potential to extend to a worldwide scale using Starrag's infrastructure. From a technology perspective Heckert surprised Starrag with its excellent competencies especially in cost-efficient engineering and manufacturing in combination with very high-quality standards. Besides these partly positive insights, Starrag also soon encountered difficulties with the labour unions; however, after some negotiations, these problems were resolved. Over the course of the next couple of years the steering committees were to be changed; however, they still exist as a matrix to the two units within the StarragHeckert group.

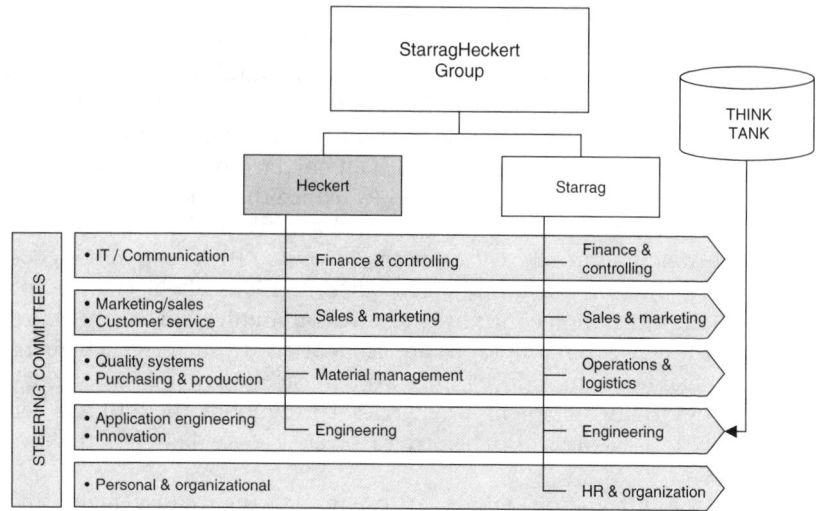

Figure 4.34 Organizational integration of Heckert with boundary-spanning structures

The Think Tank, consisting of the CEO, the managers of Starrag and Heckert, and product and R&D managers, was initially established to identify new innovation ideas but was soon transformed into the overall R&D and innovation steering committee of the whole StarragHeckert group. The Think Tank is concerned with the identification of technology-based synergies, the management of technology and product roadmaps, and the release of the major R&D project milestones. This ensures the avoidance of redundant developments and the overall commitment of the business unit and R&D managers to joint R&D projects.

The Think Tank's initial tasks, and especially those of the head of R&D, were to investigate, push or discontinue the running R&D projects at Heckert. Besides one project, the pipeline was filled with strategically very important upcoming products and thus the focus on Heckert's pipeline was strengthened rather than reduced. Furthermore, it was soon found that Heckert could use Starrag's spindle which it had formally sourced externally. This leveraging of a complete module was the first technology-based synergy to be achieved at the beginning of 1999.

Already during the acquisition in mid-1998 a highly experienced engineer and production head at Starrag had identified new upcoming market needs on the customer side, such as a smaller and cheaper five-axe milling machine. This manager soon came up with the idea of making use of Heckert's smaller machines and building on some additional axes. However, the realization of his idea was slightly postponed mainly due to the integration efforts and some hesitations from the sales side. Nevertheless, as Starrag's products

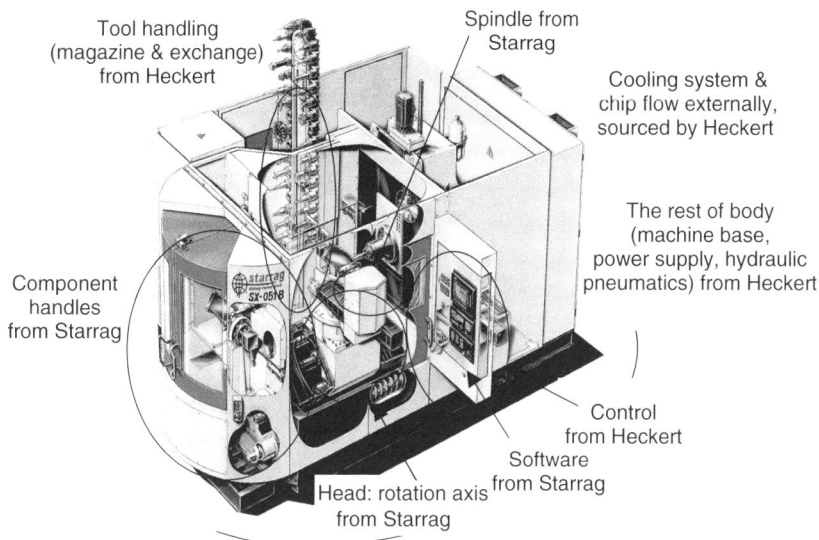

Figure 4.35 SX-051, Platform 3 developed on Heckert's base machine

needed some successors, the first joint product platform project was initiated in the summer of 1999. At approximately the same time, the CEO and head of O&L fully returned to their positions in Switzerland and handed over leadership at Chemnitz to their deputies at Heckert. The new platform product, called SX-051, was developed to serve future market demands for smaller and cheaper blade-milling machines, furthermore also referred to as platform 3 as it became the smallest platform. Instead of developing a completely new machine, Heckert's small base machine, the CWK 400 D became the core of the new machine and further modules from existing Starrag machines, such as the spindle, the rotation A- and B-axis, the component handler and also the software and related technologies, were added (see Figure 4.35).

Whereas the purchasing of the elements was still conducted by Heckert, using their good contacts, Starrag took over the assembly of the platform. In order to do so, six to seven people from assembling were sent to Heckert to learn and then to transfer the assembling knowledge. This increased the production depth at Starrag and made more efficient use of the available capacity. Even though the machine did not contain any modular innovations, the new configuration provided the market with high added value through its size and cost. The first machine was already sold in the beginning of 2000. This half-year time to market was a record for Starrag and initiated the strategic focus on platform strategies.

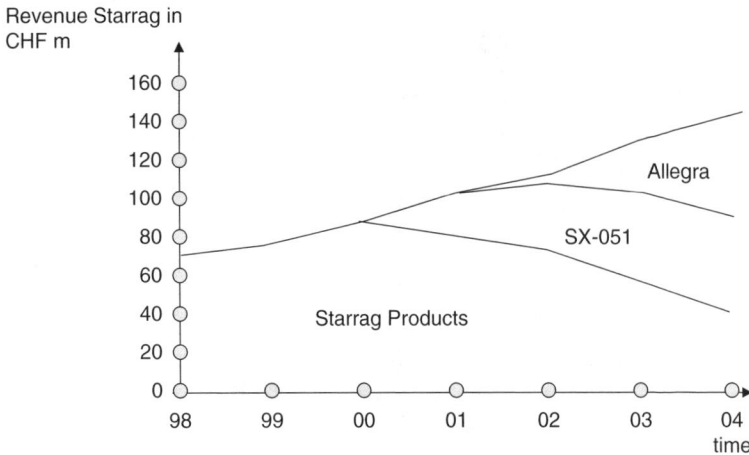

Figure 4.36 The Allegra project as the future product pipeline

Thus in the September 1999 the Think Tank, specially guided by the CEO and the head of R&D, started the Allegra project (see Figure 4.36). This platform strategy project defined several platforms which should be used for both the Heckert and Starrag machines and markets (see Figure 4.37) and thus reduce the investment cost, costs for spare parts, and increase the economies of scale. The objective was to drastically raise the innovation rate: 'in 2004 at least 50 per cent of the revenue at Starrag Rorschacherberg will be derived from products [in 2000] not yet in existence'.[18] The Allegra project was steered by the Think Tank, headed by its initiator, the head of R&D who was the most senior engineer, and his counterpart from Heckert, which was separated into the different platform groups and supported by functions such as finance and controlling, production, quality, customer service and sales. The platforms were to be built from modular building blocks from the existing and new modules and technologies (see Figure 4.38).

As a result of the newly adopted product platform strategy, a subsequent platform, PF 1, of very large size was developed. Whereas platform 3 was an adaptation of an existing machine, platform 1 was a completely new joint development project. The resulting products, HEC 1000–1600 (Heckert product line) and STC 1000–1600 (Starrag product line) were the substitutes for the older CWK 1000–1600 and ZT line (see also Figure 4.39). In the beginning the project team consisting of twelve Starrag and Heckert engineers met and collaborated to design and develop the base machine. Then the engineering of the different modules was assigned to the different teams according to the prevaling competencies. The subsequent manufacturing, by comparison, was separated according to available capacity. The project

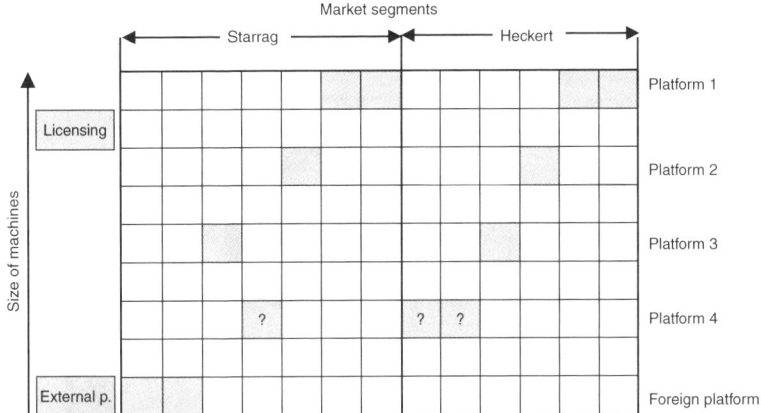

Figure 4.37 The platforms in all sizes will serve Starrag and Heckert markets

collaboration was supported by videoconferences about two to three times a week and monthly meetings often took place in Nürnberg, half way between both locations. In order to facilitate information flow, new information transfer equipment was established. Additionally, transfer of the engineering drawings was facilitated by the identical CAD systems used within the two companies. The newly developed products excelled through improved acceleration, dynamics, infeed and stiffness. These innovations were feasible through fusion of the different engineering know-how. Additionally, Heckert's capability to achieve cost-efficient designs was used to reduce the overall production cost of up to 30 or 35 per cent in comparison to that of the ZT machine. The covering of the new machine was done by an external designer, which gave the machines a new and joint StarragHeckert layout. The first prototype was finalized in June 2001.

Besides this technology integration, StarragHeckert transformed into a Holding structure in early 2000 with a joint sales organization. This, however, was organized according to the products or markets and thus the sales integration did not proceed as quickly as initially hoped.

The third product platform, PF 2, initiated in spring 2002, was developed to replace the mid-size Starrag machines (ZS-line). Similar to platform 3, it was based on an existing quite new Heckert base machine, CWK 630D, with additional modules from existing Starrag products and also external partners. For example, the platform 2 was equipped with a new torque motor from external partners. This machine, the STC 630, was also characterized by some incremental innovations.

In subsequent years, the product platform strategy was extended to a very large platform and also one very small one. It helped to replace the old products and highly increased the innovativeness of the joint companies

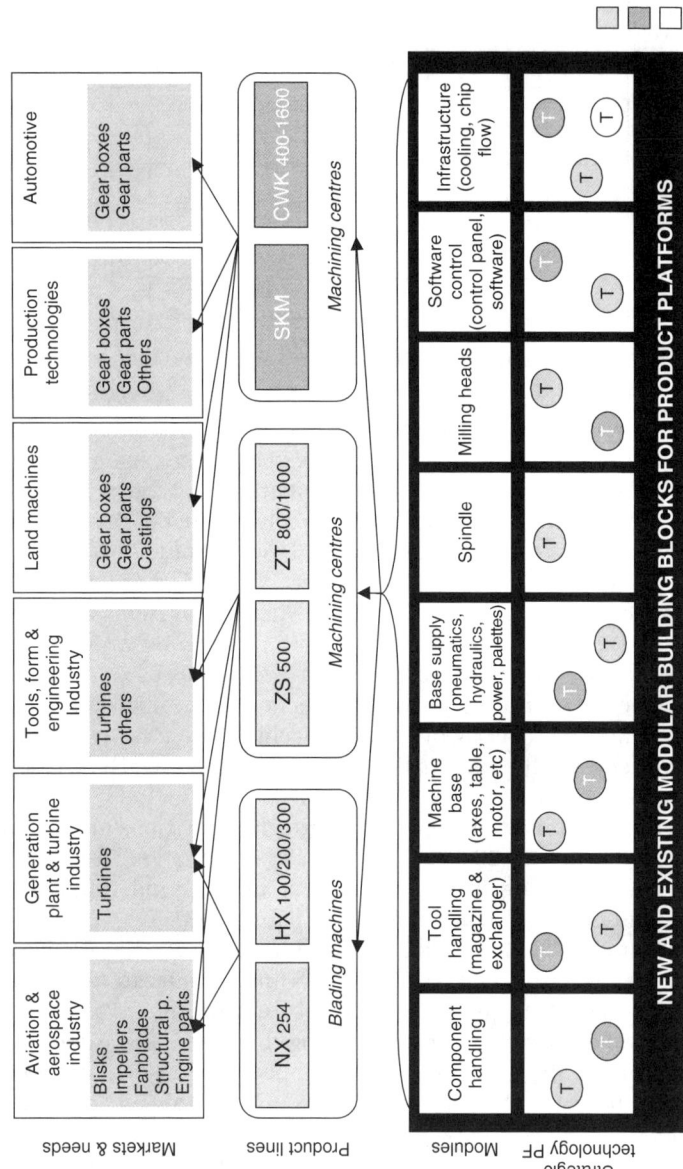

Figure 4.38 Modular building blocks for Starrag and Heckert product platforms

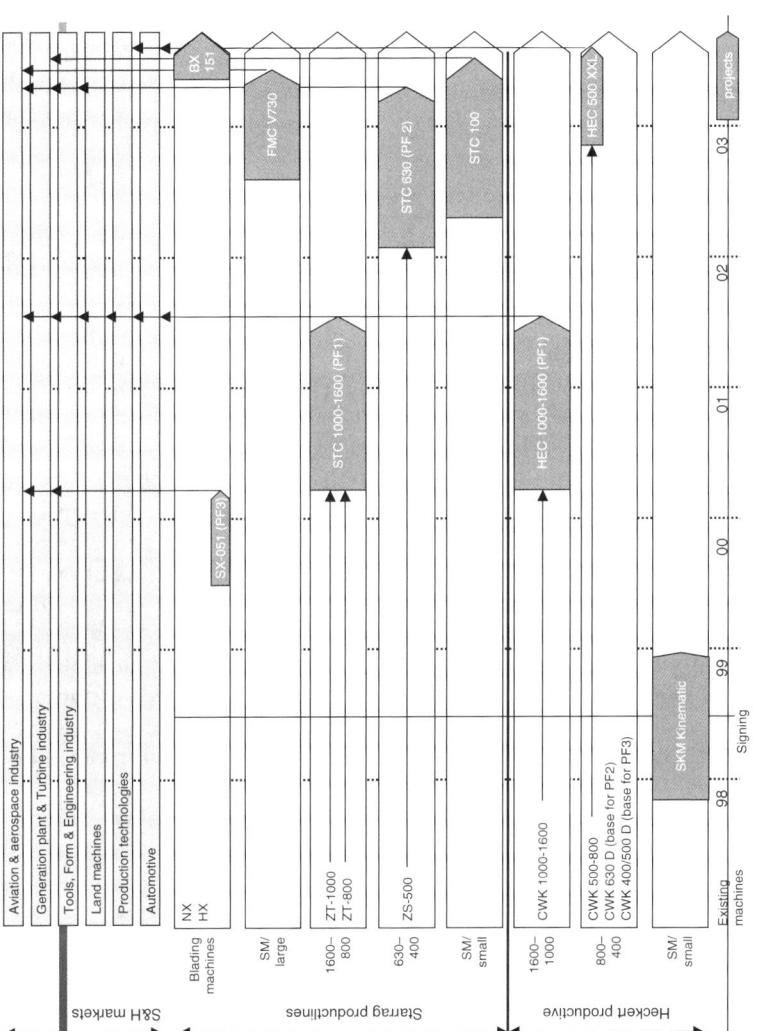

Figure 4.39 Product integration roadmap: platform strategy

(see product roadmap, Figure 4.39). Furthermore, Starrag was transformed into a manufacturer of machines for the medium- and upper-market segment with much greater awareness of cost-efficiency.

Besides the technology and product integration, the management team decided to further merge the sales forces by organizing it according to geographic regions rather than products or markets, from the beginning of 2003, in order to reap market synergies. Furthermore, SAP and other business processes, such as the innovation process would be implemented over both companies to increase the integration and depth even further to reap the increased synergy potential.

4.3.5.5 Interrelations, conclusions and discussions

Again the question arises of which aspects, such as acquisition type, initial conditions, strategy processes and external developments, influenced the technology-based value creation within Starrag's acquisition of Heckert.

Generally the *acquisition* can be described as a friendly, relatively large acquisition of supplementary markets and products and similar, partly complementary technologies within a fairly stable market environment. The technologies acquired were state-of-the-art, partly key and partly base technologies. The achieved technology-based value creation can be described as resource deployment and innovations. Similar technology platforms were consolidated and knowledge was transferred and substituted, subsequently new product platforms and product lines were developed and marketed achieving incremental, modular and architectural innovation. Additional to technology-based value creation, value creation from scale effects within procurement, sales and production were achieved.

The highly successful technology-based value creation was facilitated by the *initial conditions*, such as an excellent business unit and technology-strategic fit, similar technology platforms, and complementary and redundant process technologies such as the design-to-cost engineering capability or similar CAD systems. Furthermore, the existing products were based on a highly modular structure and thus easy to reconfigure and combine. Additionally, all competencies and technologies were sufficiently attractive and very well mastered by both partners. Additionally, the Starrag CEO's familiarity with Heckert's products and competencies facilitated the acquisition and integration process. However, probably the most important influential factor which enabled excellent technology integration was Starrag's and in particular its CEO's willingness and readiness to change Starrag itself to jointly develop into a newly merged StarragHeckert. The lack of a formalized acquisition process or M&A team did not have any significant influence. Only the different organizational cultures caused some difficulties.

The design of the *strategy processes* also greatly influenced value creation. Whereas the acquisition process was almost non-influential, subsequent actions during the integration process impacted on value creation in its stead. The strong focus on an excellent team and representatives increased the value creation potential. Furthermore, the initial very strict and direct leadership of the CEO, balanced by communication, commitment, enthusiasm, strategic workshops and the like, resulted in a joint reorientation of both companies towards one common goal. Additionally, the boundary-spanning structures, such as the Steering Committees and the Think Tank, facilitated technology integration and ensured the commitment of all parties. Last but not least the stepwise and patient approach towards technology-based value creation allowed slack for trust building and cooperation and thus resulted in an intense increase in innovation rate and a significant reduction in product development costs and time.

Also the *external developments* impacted on the technology-based value creation. The change in customers' needs and the access to an increasingly competitive medium segment increased the urge to lower costs and increase innovativeness, which resulted in the platform strategy.

4.4 Conclusion on Cases from Reality

At the beginning of this book it was explained that technology-based value creation in corporate acquisition is of utmost importance and research questions were posed which aim to improve the understanding of technology-based value creation in acquisitions and to arrive at applicable concepts for mastering technology-based value creation. In order to appropriately answer the first research question, practice was consulted by applying a case study research approach. Thus, as a first step, a theoretical framework for analysing the case studies was derived from state-of-the-art literature. This framework indicated that the acquisition type, initial conditions and the acquisition and integration process highly impact on the short- and long-term innovativeness of merging companies and their ability to redeploy their resources. In a second step, five case studies were diligently selected according to set requirements. These case studies were described according to chronological order and aspects derived from the theoretical model were discussed individually. It turned out that indeed technology-based value creation, similar to general value creation in acquisitions, is dependent on acquisition type, initial conditions, strategy processes and additionally on external developments. These findings will be compared in a comprehensive way in the next chapter. Furthermore, the findings will be complemented and challenged by existing research. In this way a new comprehensive understanding on which aspects determine successful technology-based value creation after acquisitions have taken place will be formed. Subsequently this new understanding will be incorporated into a technology-based strategic acquisition and integration management concept, introduced in Chapter 6.

5
Model of Reality: A New Understanding of Technology-based Value Creation in Corporate Acquisitions

Within the last chapter, five different real-life cases of acquisitions where technology-based value creation played a key role were analysed. Within this present chapter, these cases will be compared in a cross-case comparison and a new and holistic understanding of aspects relevant to achieve technology-based value creation will be derived. As the number of cases is limited and the variables compared are quite broadly defined, the derived conclusions and understanding are not quantitatively, empirically profound but they represent a good overview of the interrelations and impact of the different variables and thus contribute to a holistic understanding of corporate reality. This is in line with the pursued exploratory research approach (see Chapter 1.4). In addition the findings are complemented by research results from state-of-the-art theory. With the derived model of understanding, the first research question: 'Which aspects in corporate acquisitions in innovation-driven industries determine the successful realization of technology-based value creation?' will be answered. Based on this improved understanding the concept for a technology-based strategic acquisition and integration management will be developed in the subsequent chapter.

The following sections which describe the model of reality (see Figure 5.1) will initially address the observed types and mechanisms of technology-based value creation and their occurrence in the course of the acquisition integration. Subsequently, it will be outlined in which acquisition types which technology-based value creation becomes critical. Thus the different challenges in acquisitions which aim for market access and those which target new competencies are inferred. Next, the characteristics of target and acquirer which are required to enable technology-based value creation, the initial conditions, are summarized and discussed. Furthermore, an explanation of the impact of strategy processes on technology-based value creation will be provided. The process design and particularly the fulfilment of

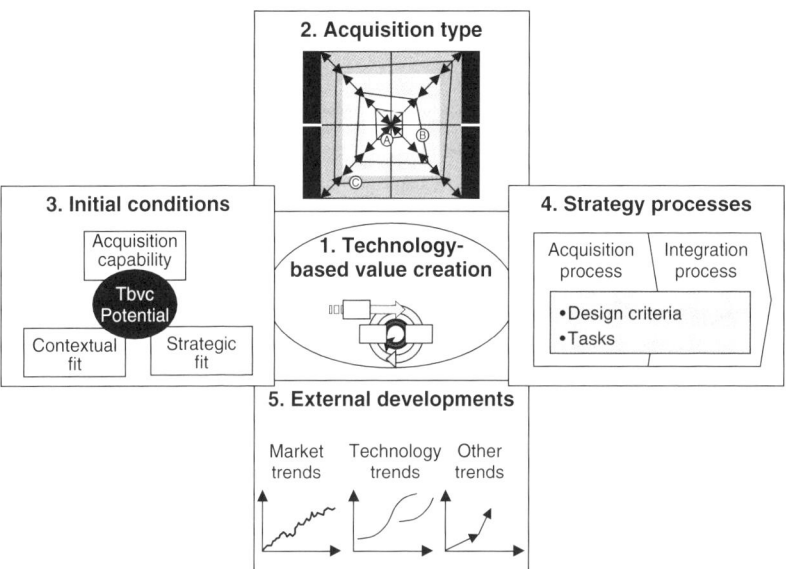

Figure 5.1 Elements of the model of understanding

various tasks during acquisition and integration significantly determine the likelihood of successful technology-based value creation. Last, but not least, the additionally observed impact of external developments will be mentioned. Thus this chapter elaborates on the key success factors for achieving increased innovativeness and an efficient resource deployment following acquisitions.

5.1 Technology-based Value Creation – Quick Gains and Long-term Success

Within all cases a certain level of technology-based value creation was achieved. In order to explain this, the model on innovation and resource deployments with the underlying resource integration mechanisms is referred to as explained in Chapter 2.1.

5.1.1 Innovation and resource deployment

Generally both *innovation* and *resource deployments* were observed in most of the acquisitions. The observed innovation mechanisms were mostly innovations based on the leveraging of technologies, modules and products into new products and markets, or the reconfiguration of technologies and modules, into new product platforms. An example is the integration of various modules such as platform and packaging into the newly developed

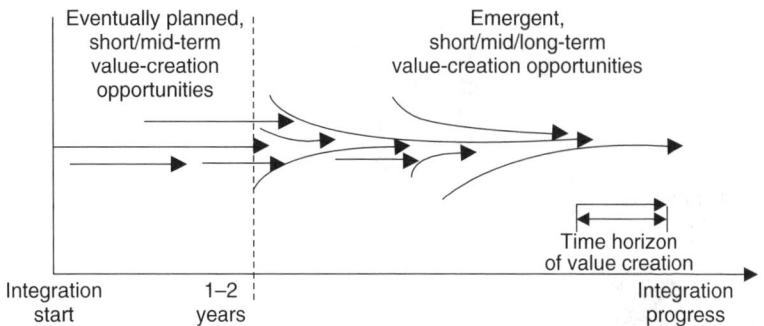

Figure 5.2 Planned and emergent technology-based value-creation opportunities

Unitron and Phonak products or the development of StarragHeckert's product platforms. The observed resource deployment was the transfer or substitution of technologies and knowledge which resulted in improved competencies or lower production costs, for instance the transfer of manufacturing infrastructure. For a detailed overview of the different resource integration mechanisms see Figure 2.1 in Chapter 2.

Furthermore, the technology-based value-creation opportunities varied in their *time horizon*. Whereas some of these technological synergies had only a very short-term effect, such as the unique integration of technologies within the product development process,[1] others, especially the resource deployment mechanisms or platform development projects, promoted long-term value creation of the merged companies.

Another phenomenon was the *emergence* of technology-based value-creation opportunities. In various cases, the technology-based value creation was assigned limited priority during the acquisition as its impact seemed insignificant. However, in the course of acquisition integration, especially tight integration, technology-based value-creation opportunities emerged and often became the basis for long-term success (see Figure 5.2). For example, Phonak did not even consider technological synergies at the time of acquisition and yet today most of the products and their increased innovation rate are based on the possibilities of leveraging modules and technologies. Besides the observation of the emergence of technology-based value-creation opportunities, a certain *pattern of occurrence* of different resource-integration mechanisms in the course of the acquisition integration progress was identified.

5.1.2 Quick gains and long-term successes

As was observed in the Unitron and Heckert acquisition, the technology-based value creation often starts with quick gains such as easy-to-achieve innovations. Thus within the first integration year, often complementary and previously already developed modules are leveraged into the first joint

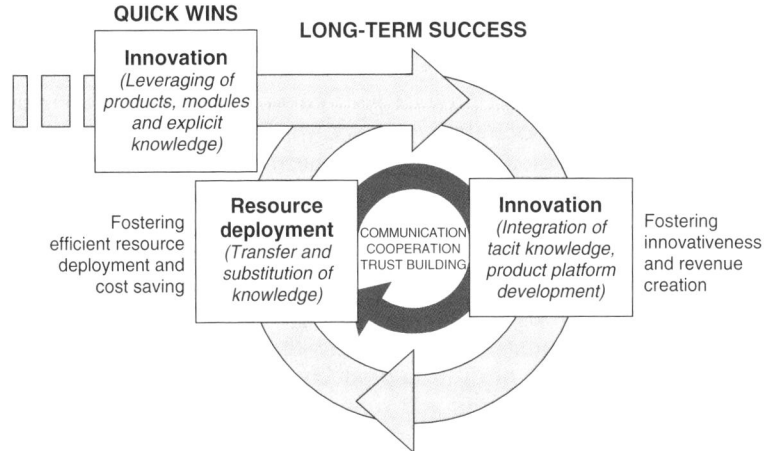

Figure 5.3 Wheel of technology-based value creation

product developments or sold on other markets – often referred to as cross-selling opportunities. This integration of highly explicit and easy-to-combine modules often resulted in quick wins, as the integration mechanisms required solely the detailed definition of product interfaces, little cooperation or trust and a limited amount of time. For example, shortly after acquisition Heckert integrated the spindle from Starrag, while Phonak developed a new product with the core modules of Unitron within half a year. As these quick wins often do not require tight organizational integration and thus facilitate the retention of key engineers[2] and the innovativeness[3] of the acquired company, this approach often seems appropriate for the acquisition of technology-intensive companies. However, it was further observed that this short-term-oriented approach does not lead to long-term success. After reaping these quick gains, the realization of further technology-based value-creation opportunities becomes much more complicated and requires a different integration approach[4] (see Figure 5.3).

Additionally, it was observed that companies from this point on[5] can only successfully realize technological synergies if *resource deployment* and *innovations run in parallel* and promote each other. Thus after the quite straightforward integration of modules or products, other innovation mechanisms occur. Then not only products but also technologies and more tacit knowledge are integrated into the joint module and product developments. Furthermore, as was observed within the Plasma-Therm and Heckert acquisitions, product platforms and thus the reconfiguration of technologies is pursued. These innovation mechanisms, however, cannot be achieved with low organizational integration or the mere definition of interfaces. The joint development projects require cooperation, communication and trust building. Furthermore, whereas the integration of modules into new products does not require learning about the underlying technologies, the integration of tacit knowledge or the development of product platforms require an indepth mutual and technological understanding.[6] The knowledge flow or learning is fostered and enabled by resource-deployment mechanisms. Thus the advanced usage of tacit knowledge to achieve innovations requires knowledge transfer and eventually also knowledge substitution and thus resource deployment. For example, the successful development of a new product platform and line for the magneto business of Plasma-Therm required the transfer and co-location of the research group from Alzenau to St Petersburg. Another example is the transfer of packaging competence from Phonak to Unitron. In order to foster the innovativeness of Unitron products, Phonak had to transfer CAD and, engineering competencies and substitute packaging competencies at Unitron. Thus this learning and transferring of knowledge, always conduced via cooperation, communication, co-location and the like, builds the basis for innovativeness. Thus technology-based value creation gets into a cycle of innovation-fostering resource deployment and innovations which again require resource deployment (see Figure 5.3). This self-accelerating cycle fosters the occurrence of emergent technology-based value-creation opportunities.

It is well known that this evolvement of technology-based value creation does not occur automatically. Thus the question arises of how this ideal typical occurrence of planned and emergent innovation and resource deployment opportunities is related to the different acquisition types, is dependent on the company's specific characteristics, is determined by the design and tasks fulfilled during the strategic acquisition and integration processes, and is influenced by external developments.

5.2 Acquisition Type – Indeed not all M&As are Alike – and that Matters

Joseph Bower has written a famous article: 'Not all M&As are alike – and that matters'[7] and made an exciting speech: 'When we study M&A, what are we learning?'[8] on the diversity of corporate acquisitions and researchers' lack of its consideration in their attempt to explain reality. His opinion is shared by various recent researchers.[9] This book attempts to take this claim into account and indeed results show that M&As are really not alike and that matters also for technology-based value creation.

Generally researchers have observed so far that friendly acquisitions and those which are associated with only small debt are more likely to generate technology-based value creation. This is due to the facts that hostile take-overs do not include a due diligence, and thus lack an evaluation of the technologies, and that a negative climate between the two companies has a negative impact on the cooperation and collaboration environment required for technology-based value creation.[10] Furthermore, it was observed that high levels of debt decrease managers' willingness to take risks and thus have a negative effect on R&D investments.[11]

Besides these general characteristics impacting on technology-based value creation, the different acquisition types also influence innovation and resource deployment mechanisms. The distinction between various acquisition types made within this book is particularly important as technology aspects are usually only considered within the context of innovation-driven acquisitions and hardly mentioned in acquisitions that aim for the development of new businesses or for market access. Thus practitioners here can infer how technology related challenges specifically apply in a specific company and acquisition situation.

The main challenge was to apply or develop a categorization of acquisitions which is easy to understand, applies to most acquisitions and can be used to discuss the various types and associated mechanisms of technology-based value creation.

Generally, three different archetypes of *acquisition strategies* and related *acquisition characteristics* can be identified. These are referred to as:

- **Venturing acquisition:** aiming for access to new and unfamiliar technologies which can be the base for a new, eventually discontinuous business;
- **Substrate for growth acquisition:** aiming for acquisition and internalization of complementary or additional competencies and to build up new businesses to take advantage of industry developments and foster company growth;
- **Play for scale acquisition:** aiming for access to complementary markets or geographic scope and the realization of scale effects.

It is important to mention that these acquisition types represent a rough categorization of acquisition strategies and characteristics. Individual acquisitions may not be allocated exactly to one of these three types but may be a mix of them.

Before the three acquisition types will be related to technology-based value creation, their typical characteristics are outlined as these finally determine the occurrence of technology-based value-creation opportunities.

5.2.1 Acquisition characteristics

The acquisition types have four characteristics (see Figure 5.4):

- business-relatedness
- relative size
- technology development
- market dynamics.

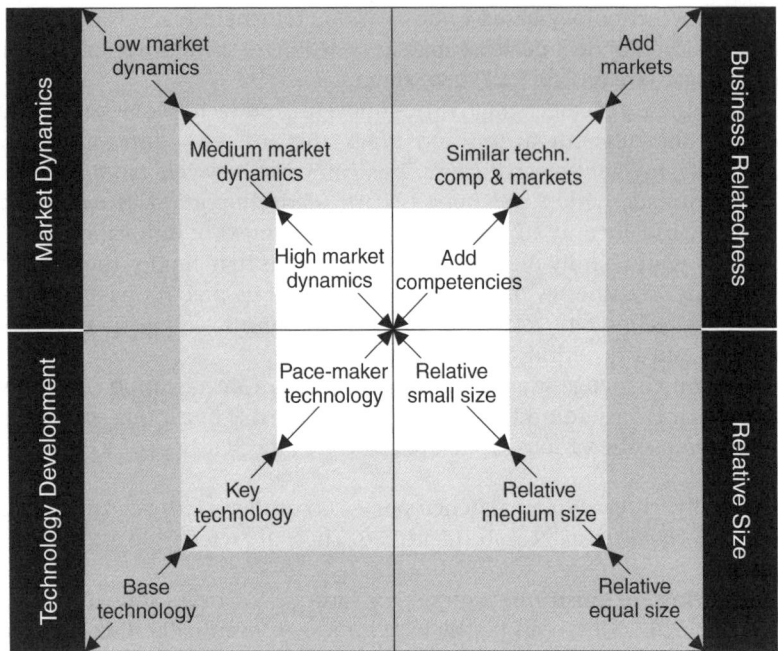

Figure 5.4 Characteristics of the acquisition types

The *business-relatedness* of acquirer to target indicates the overlap of their competencies and markets. It determines whether the technology-based value creation focuses rather on the integration of complementary additional technologies to foster innovativeness or if the consolidation and efficient deployment of redundant or similar resources is in the fore. The different *business-relatedness* options and the associated emergence of technology-based value-creation opportunities are outlined in Figure 5.5.

Generally Hoskisson and Hitt (1988) and Baysinger and Hoskisson (1989) have found a positive impact of *business-relatedness* on R&D performance. Similar conclusions have been made by Hagedoorn and Duysters (2002) who have found a positive impact of related product markets on the innovation potential of the merged companies. Figure 5.5 shows more specifically that staying in a similar technology and market field mainly offers opportunities for cost reduction via an efficient resource deployment. Entering new markets offer great cross-selling opportunities and chances to efficiently deploy similar competencies, whereas entering new technological fields offer the chance for several innovation mechanisms such as leveraging or reconfiguring knowledge bases to create new product platforms or lines.

	Existing markets	New markets
Existing competencies	Efficiently deploying resources, integrating new technologies	Efficiently deploying redundant resources
New competencies	Integrating and reconfiguring complementary and supplementary resources	Little resource integration and transfer

Figure 5.5 Different resource-integration mechanisms for different business relations

The second main characteristic of the different acquisition types is the *relative size* of target to acquirer. The majority of researchers have found a positive impact of a large relative size of target to acquirer,[12] Ernst and Vitt (2000) have found that fluctuation of innovators and innovation performance reductions are less likely with smaller targets. This contradiction needs clarification. Thus it can be shown that different relative sizes of targets pose different challenges for the achievement of technology-based value creation.

Within the case studies it was observed that acquisition of relatively small, often highly innovative targets often offers opportunities for quick gains, such as immediate integration of acquired products or modules, which do not require much integration effort. Thus innovators are encouraged to work independently and to initially stay with the company. The subsequent value-creation opportunities, however, become more difficult to achieve. Despite the fact that smaller targets due to their size are mostly comparatively easy or less complex to integrate from a content perspective, the differences in corporate contexts[13] between the large acquirer and the smaller firm often cause difficulties in intensifying technology-based value creation. Quite often the tighter integration of a small target, which is required for these types of long-term technology-based value creation, destroys the context of the target company and thus results in a destruction of the underlying competencies. This destruction becomes apparent with the loss of innovativeness or the later departure of key innovators and other personnel. Thus at the acquisition of smaller companies long-term technology-based value creation is difficult to achieve and often not achieved at all due to contextual differences resulting in a destruction of the value of the target through the loss of its knowledge base and innovativeness. For example, within both cases of relatively small acquisitions (Hilti and Fillpack) contextual differences were described as difficult to master. Furthermore, within both cases, long-term value creation was not achieved.

Technology-based value creation at the acquisition of relatively large targets occurs quite differently. These acquisitions often provide only limited

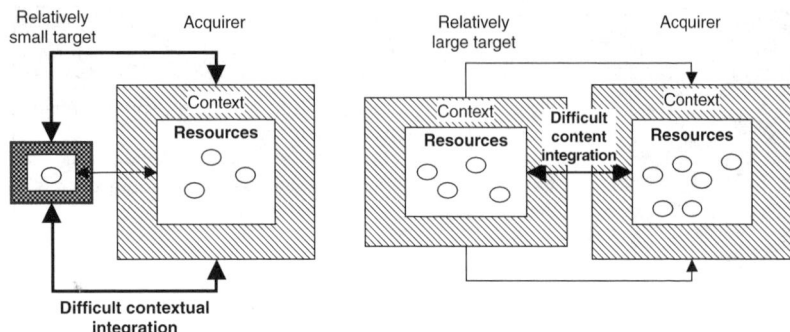

Figure 5.6 Organizational and contextual integration for relatively small and large acquisitions

opportunities for quick gains such as the immediate leveraging of modules or new technologies within the first months of the integration; however, they comprise tremendous subsequent value-creation opportunities such as the efficient deployment of capabilities within the companies and the development of platform products. Whereas in smaller acquisitions the required, contentwise, integration effort is limited and contextual adaptation poses the main threat, in large acquisitions the opposite occurs. Due to often quite similar organizational contexts, there is less risk of competence destruction or innovators leaving; however, the integration effort to consolidate what the companies are doing is very demanding. Good examples are the integration of the Plasma-Therm, Heckert and Unitron acquisition targets. Thus technology-based value creation requires tight and challenging technology integration accompanied by the adoption of the often quite similar organizational contexts. The differences in contentwise and contextual integration at relatively small and large acquisitions can be seen in Figure 5.6.

The third main characteristic of the different acquisition types is *technology development*. This characteristic is somewhat related to the relative size of the company. Generally, it can be distinguished whether the main technologies underlying technology-based value creation are highly mature or very young. In order to illustrate the dependence of different resource integration mechanisms on technology development, the main technology development-related curves are illustrated in Figure 5.7. These curves show technology development with the S-curve (Foster, 1986) and the related market dynamics such as the development of the innovation rate driven by the technology development (Abernathy & Utterback, 1978) and the technological mobility within the market.

Acquisitions of companies with pace-maker technologies which predict a discontinuous market development mainly aim to learn about the new technology without expecting immediate returns from the acquisition.

Model of Reality   109

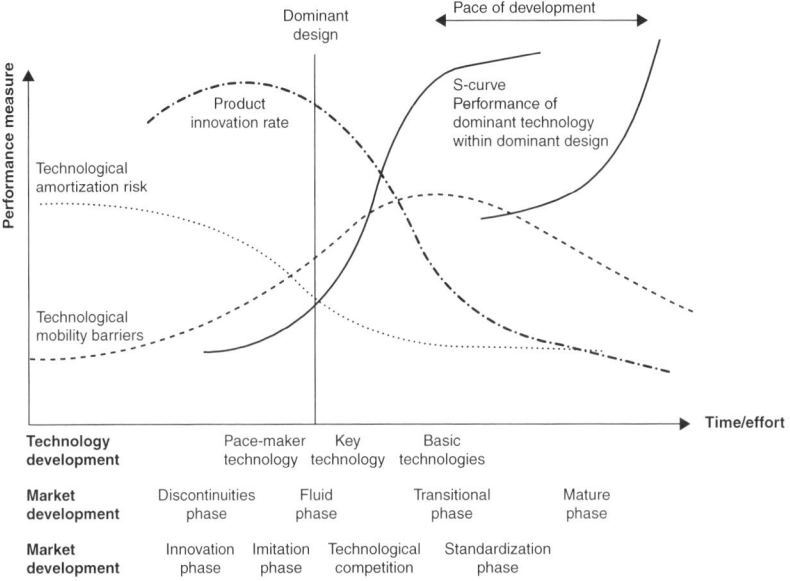

Figure 5.7   Technology development and related curves
Source: Foster (1986: 31), Abernathy and Utterback (1978),
Mueller and Tilton (1969: 578).

Often these types of acquisitions are carried out by venture units of larger companies. As no dominant design is developed by that time and innovativeness is intense and indicates in various directions, these acquisitions are associated with very high risks and are often regarded as venture projects. Thus detailed technology-based value-creation opportunities cannot be anticipated. The acquisition of more mature technologies after the dominant design provide potential for technology-based value creation. The younger the technology the more likely the potential for innovation, which is visualized by the high innovation rate associated with the occurrence of the dominant design. Mature technologies can be utilized by redeploying them or achieving architectural or incremental innovations. This fits with the associated industry dynamics, as Abernathy and Utterback (1978) have shown. Competitiveness in markets around a mature technology is mostly determined by economies of scale and cost-competitiveness. Following the categorization of Mueller and Tilton (1969) it can furthermore be argued that the acquisition of non pace-maker technologies aims to overcome technology mobility barriers and thus allows the acquirer to enter new businesses.

Last but not least, *industry dynamics* impact on technology-based value creation. In very fast industries, where product life-cycles are extremely short, acquisitions are often conducted with only the objective of providing the next-generation product. The long-term perspective is not as important as the requirements from the market which drastically change over time and thus need immediate and adaptive reactions. Thus technology-based value creation in very fast industries is often limited to the realization of the first quick gains such as the integration of the target's products into own markets. The following value-creation opportunities are often not important for the competitiveness of the acquirer and thus the target is either divested or completely absorbed and redistributed within the company. In less pacing markets all sorts of technology-based value creation, short- and long-term opportunities, are possible and essential.

5.2.2 Acquisition types and strategies

After having introduced the different characteristics of acquisitions and their impact on technology-based value creation, these characteristics can be linked to the introduced acquisition strategies. This interrelation between the mentioned acquisition strategies and described characteristics is visualized in Figure 5.8. From this the descriptions of the acquisition types and the possible technology-based value-creation opportunities and mechanisms will be found in Figure 5.9.

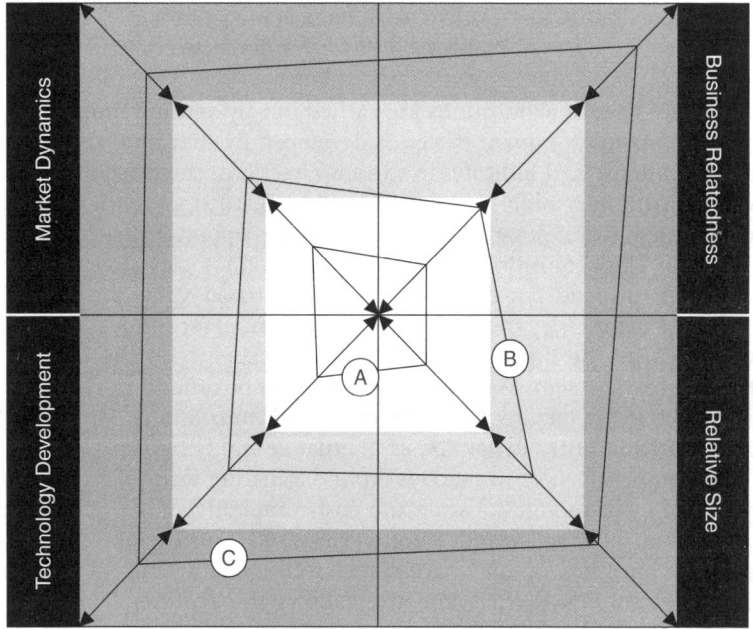

Figure 5.8 Characteristics of the three acquisition types: Ⓐ venturing acquisitions, Ⓑ substrate-for-growth acquisitions, Ⓒ play-for-scale acquisitions

Nr.	Acquisition type	Acquisition characteristics	Technology-based value creation	Cases
A	Venturing acquisition	• New, unfamiliar competencies • Relative small size • Pace-maker technologies • High market dynamics	• Resource integration • Modular innovation & add revenue • Mostly low-hanging fruits • Difficult long-term tbvc • Short-term focus	**Hilti/ Ammann LT**
B	Substrate for growth acquisition	• New, familiar competencies • Medium relative size • Key technologies • High/medium-market dynamics	• Resource integration, reconfiguration, transfer, substitution, … • Cost saving & revenue increase • Low hanging fruits & long-term tbvc • Short- and long-term focus	**Fillpack/ Aseptofill** **Unaxis/ Plasma-Term**
C	Play for scale acquisition	• Similar, redundant competencies • Relative large size • Key or base technologies • Medium or low-market dynamics	• Resource transfer & substitution & reconfiguration • Cost savings, product platforms • Little low-hanging fruits • Long-term focus	**Phonak/ Unitron** **Starrag/ Heckert**

Figure 5.9 Summary of the acquisition types and characteristics

(1) Venturing acquisition: The acquisition strategy of venturing acquisitions is to gain access to new and unfamiliar technologies which can be the base for new and discontinuous businesses. These acquisitions are mostly characterized by the sourcing of new competencies, a relative small size and pace-maker technologies and they happen often in a quite dynamic market. From the argument above it can thus be inferred that within such acquisitions the quick gains from easy-to-achieve innovation opportunities are mostly in the fore. The target offers the opportunity for attractive short-term innovation possibilities but is difficult to be integrated further to achieve long-term success due to the different corporate contexts. Furthermore, these acquisitions often do not even aim for long-term contribution as the industry dynamic is too fast and thus requires short-term thinking. Hilti's acquisition of Ammann LT was an example.

(2) Substrate for growth acquisition: The acquisition strategy is to acquire and internalize complementary or additional competencies and to build up new businesses to take advantage of industry developments and to foster company growth. Thus the acquisitions are mostly characterized as the acquisition of new, however familiar, competencies within related markets, often by a medium relative size and pace-maker or key technologies within fast- or medium-paced markets. The associated technology-based value creation thus offers a variety of opportunities. Quick gains as well as long-term technology-based value creation can be achieved facing contentwise and contextual integration challenges. Furthermore, technology development and relatedness allow for modular as well as architectural innovation and also resource deployment. Market dynamics demand medium- and long-term focus on the acquisition and associated value-creation opportunities.

(3) Play-for-scale acquisition: The play-for-scale acquisitions mostly aim for achieving a dominant market position through gaining access to additional markets and reaping scale effects. The acquisitions can be characterized as the acquisition of additional markets with highly similar technologies, a relatively large-sized target, key or base technologies and often within lower market dynamics. The technology-based value creation is mainly rooted in long-term-oriented innovation and resource deployment and is less based on quick gains. Furthermore, the contentwise integration required for these value-creation opportunities are the highest challenges whereas contextual integration aspects are mostly not as critical. Due to the integration of base technologies within fairly dynamic markets, the long-term value creation potential is of critical importance.

It can be summarized that acquisitions can be categorized into three main acquisition types with their associated acquisition strategy and characteristics. Within all of these types, technology-based value creation plays a different role and faces different challenges. Thus the acquisition types have an impact on possible value creation mechanisms and thus also on required initial conditions and strategy processes. Thus the effects of the different acquisition types will be mentioned throughout the further discussion on the aspects that impact on technology-based value creation.

5.3 Initial Conditions – Necessary, However Insufficient, Conditions

So far the essence of technology-based value creation and its dependence on the various acquisition types have been outlined. Besides these factors on how and where technology-based value creation occurs, now more specific aspects which impact on the successful realization of technological synergies are outlined. Within this section the required characteristics of the acquirer and target which enable joint innovativeness and successful resource deployment are outlined. Four different areas of these initial conditions can be distinguished (see Figure 5.10). First, technological synergies are dependent on an inherent *technology-based value creation potential*.[14] Secondly, the acquirer has to master a certain *acquisition capability* in order to be able to achieve technological synergies within the acquisition-specific context. Thirdly, a *strategic fit* between the two companies is the basis for successful value creation and fourthly, a certain *fit between the corporate contexts* of the two companies is conducive to technology-based value creation.

5.3.1 Technology-based value creation potential

The investigation of the cases was specifically focused on identifying the technology-related aspects which are conducive to technological synergies.

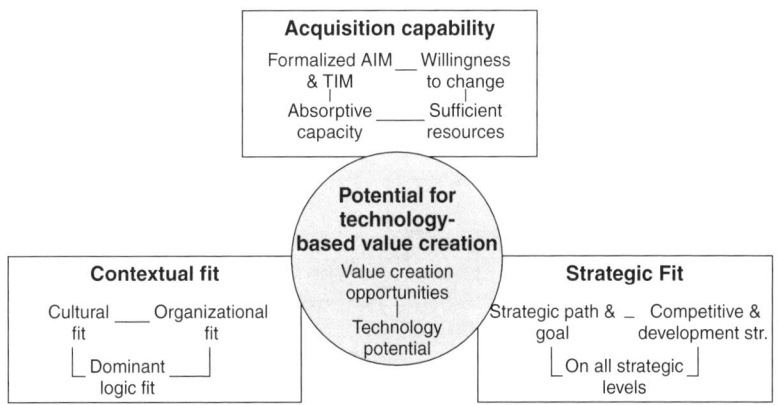

Figure 5.10 Necessary, however insufficient, conditions

Two specific conditions were identified: First of all, successful joint innovativeness is dependent on the *characteristics of the technology bases* of the target and the acquirer, and, secondly, it is determined by the *opportunities* to combine the technology bases to *create value*. Without the first aspects – a strong technology base – value-creation opportunities do not reap sufficient returns and without the latter aspects – the attractive opportunities – the best resource bases become useless. These two factors, the resource bases and the value-creation opportunities, are summarized under the technology-based value creation potential. This is defined as the *sum of attractive and realizable planned and emergent technology-based value-creation opportunities within the acquisition and integration-specific contexts.*

5.3.1.1 Characteristics of the resource bases to foster technological synergies

Within the case studies an investigation into which characteristics of the resource bases of the target and acquirer and of the achieved value-creation opportunities lead to success was undertaken. The findings were compared with and complemented by the main findings from literature on technology and innovation management and can be summarized as in the following text and in Figure 5.11.

The resource base comprises product as well as process technologies and thus also includes organizational competencies. Besides the fact that rather large technology bases are conducive to value creation,[15] three main

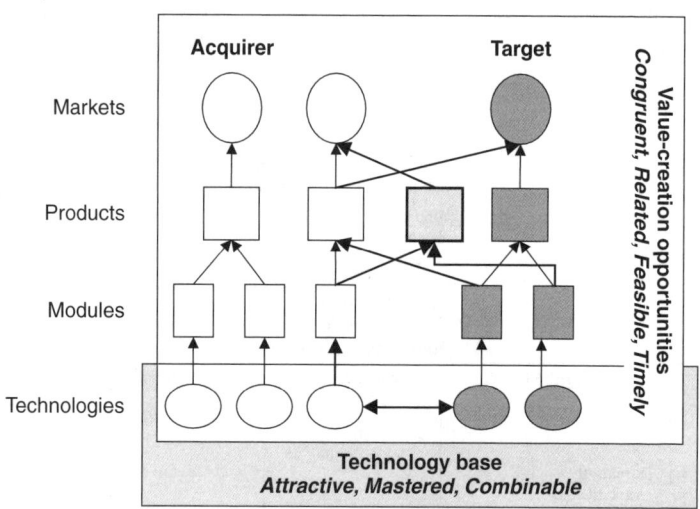

Figure 5.11 Technology-based value-creation potential

characteristics of the resource bases impacting on the *potential for technology-based value creation* were identified:

- Attractiveness of the technology base
- Level of mastering the technology base
- Ability to combine the technology base.

The first characteristics, the *attractiveness of the technology base*, indicates that technology-based value creation was only achieved if the underlying technologies were either fairly attractive or were the core technologies of the companies. For example, the attractiveness of the aseptic filling of PET bottles, one technology of Aseptofill acquired by Fillpack, was partially reduced by its dependence on barrier technologies for PET bottles. Thus the technology-based value-creation opportunities achieved limited returns. According to literature, technology attractiveness can be described by the functional scope, maturity, independence on other technologies, scalability, etc. of the technology.[16] Thus it is claimed that successful emergent and planned technology-based value-creation opportunities should be based on attractive and core technologies.

The second characteristic, the *level of mastering the technology base*, has often impacted the potential for technology-based value-creation opportunities within the acquisition cases. A poor level of mastering technologies delays developments and results in inferior and malfunctioning products. For example, the medium level of mastering the aseptic filling technology by Aseptofill or the problems in mastering the DSP factory platform by the spin-off of Unitron delayed product development and reduced the returns on launched products. This level of mastery can be indicated by experience within the technology field, number of experts focusing on that topic, infrastructure supporting technology development and also underlying intellectual property. Thus the lack of patents related to a technology can be seen as a low level of mastering the technology.[17]

The third characteristic of the technology base impacting on technology-based value creation is the *ability to combine the technology bases*. This ability is dependent on the characteristics of the resources and of their relatedness.

Whereas highly combinable technologies facilitate technology integration[18] and thus technology-based value creation, these technologies are often characterized by their explicit nature, context independence, low dispersion, modular nature or standardization. Thus these technologies are easy to imitate, substitutable and seldom rare and thus represent a lower value than highly tacit, dispersed and context-dependent resources (Barney, 1991). Thus the integration of easy-to-combine resources with lower value provides a limited potential for value creation. On the contrary, the highly challenging integration of context dependent, dispersed,[19] or tacit resources which are mostly highly valued provide the potential for sustained

116 *Mastering the Acquirer's Innovation Dilemma*

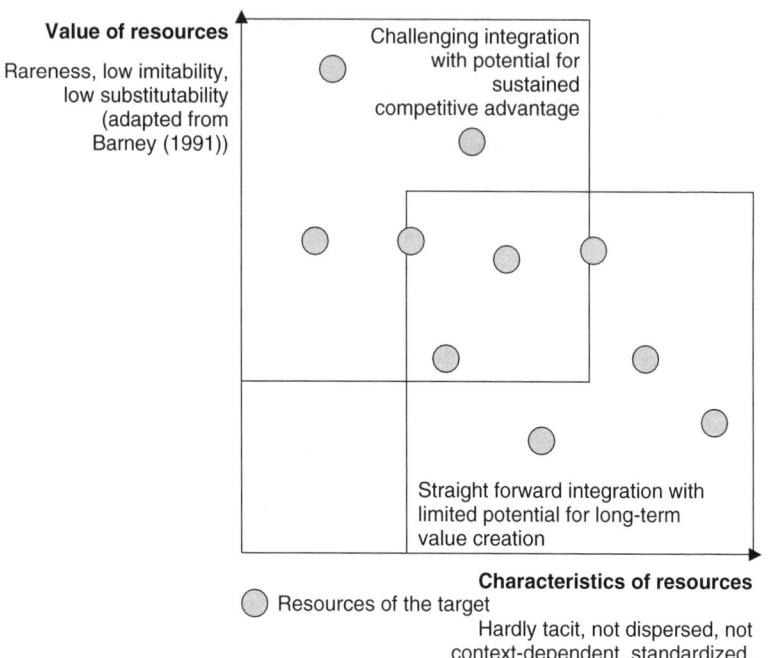

Figure 5.12 Attributes of the resources to be combined and associated value-creation potential

competitive advantage (see Figure 5.12). This was observed with the ease in leveraging the spindle between Starrag and Heckert which resulted in only a small value creation, whereas the integration of the highly tacit and context-dependent knowledge of the former owner of Ammann LT at the Hilti acquisition brought very high value creation, albert that it was very difficult to achieve.

Besides the impact of the characteristics and value of the resources on technological synergies, the relation of the resources of the target and the acquirer to each other also influences the ability to combine the resource bases and thus the occurrence of technology-based value creation. The results from the state of the art in literature on this topic are so far quite inconclusive. Some researchers[20] could not find any significance, whereas others have found that a moderate degree of overlap has a positive impact.[21] Hitt and his colleagues (1998b) and others argue that complementary resources are conducive to technological performance. Harrison *et al.* (1991) add that differences in technological resources significantly contribute to the performance of the merged firms. Ahuja and Katila (2001) argue there is a non-linear impact of the relatedness of the knowledge bases on innovation output, whereas 'high levels of relatedness and unrelatedness both prove inferior to acquiring firms with moderate levels of relatedness' (Ahuja & Katila, 2001: 215). Ernst and Vitt (2000) report that a high degree of technological proximity leads to lower fluctuation rates. These diverse statements can be better understood when distinguishing more diligently between technological relatedness and then explaining their combinability.

Whereas resource deployment can take place only between redundant, similar or complementary resources, innovations can be based on all the various relations between resources. For instance, the combination opportunities based on supplementary or unrelated resources can lead to modular innovations, while the substitutive or enhancing leveraging or reconfiguration of similar, redundant or complementary resources mostly results in incremental or architectural innovation (see Figure 5.13).

5.3.1.2 Characteristics of the value-creation opportunities to foster innovativeness

Besides a strong technology base, the technology-based value creation potential is also determined by *attractive value-creation opportunities*, which represent the possibilities of successfully commercializing technology bases, by means of innovation and resource-deployment opportunities. The attractiveness of a value creation opportunity can be measured by its net present value (NPV) or return on investment (ROI). This financial success of value-creation opportunities is determined by the main factors impacting on successful resource combinations. Within the case studies it was observed that resource combinations which matched customer needs (like the product platforms of StarragHeckert) were attributed by a certain degree of newness (such as the new packaging of the Unitron products), were timely to the market (which was the main issue with the new digital Phonak product), were developed with sufficient development competence (this was the reason to acquire Ammann LT) and did not lack manufacturing competence (which Hilti had to integrate by cooperating with Jenoptik) became financial successes. These factors are similar to the innovation criteria proposed by the literature. Whereas several authors[22] have pursued different approaches to categorize the key success factors for technology-based value-creation opportunities, within this book the following four different clusters are formed:[23]

(1) Congruence of the value-creation opportunity
(2) Relatedness of the value-creation opportunity
(3) Feasibility of the value-creation opportunity
(4) Timely occurrence of the value-creation opportunity.

A technological synergy has to be *congruent*. With congruence the inherent and consistent link between the customers and their needs and the provided

Relation	Complementary	Supplementary	Redundant	Unrelated
Description	A complementary technology enhances or improves the existing functionality of another technology	A supplementary technology provides an additional functionality to the existing technology	A redundant technology is similar to an existing one or fulfills a similar functionality	An unrelated technology provides a different functionality
Innovation	Modular, architectural, incremental	Architectural, modular	Incremental, modular	Radical, architectural
Relation (Effect-function-technology)	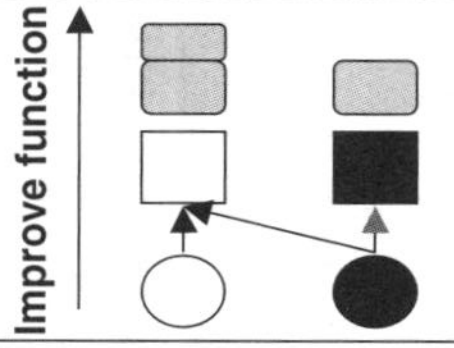 Improve function	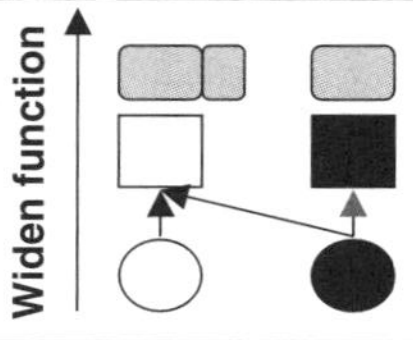 Widen function	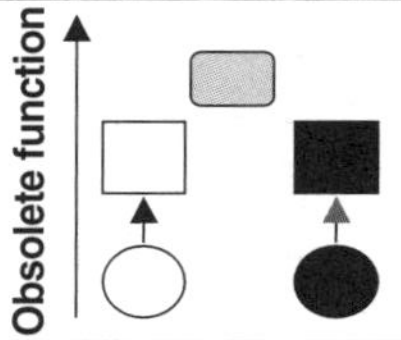 Obsolete function	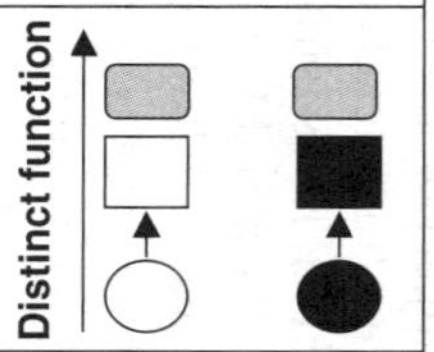 Distinct function
Resource integration mechanisms	Integration, reconfiguration, transfer	Integration, reconfiguration	Transfer, substitution	Fusion

Figure 5.13 Resource relation and associated technology-based value creation

functionality are described. This functional match of a congruent technology synergy is characterized by a balance between market pull and technology push and thus a certain degree of newness. Furthermore, congruence is achieved if all resources such as technologies and modules combined within an innovation are sufficiently well mastered. Thus a technology synergy should be based only on existing and profound knowledge[24] and be supported by strong patent coverage.[25]

A technological synergy has to be *related*. In literature and within the cases it was observed that successful innovations of a company are mostly related to previous products, markets or technologies. Attempting to achieve innovations or resource deployments in a highly unrelated field is highly uncertain.

A technological synergy has to be *feasible* with the underlying resources. An innovation project requires sufficient development knowledge and capabilities such as experimenting know-how, the process technologies and infrastructure and the skills of the participants. Additionally, the realization of the innovation requires sufficient manufacturing resources. For example, when Starrag acquired Heckert, Starrag acquired not only certain products and platforms to develop new products for a lower-range segment, but also the design-to-cost engineering competence. Without this, the innovation would not have succeeded. This required engineering or innovation competence also includes marketing knowledge and the like.

A technological synergy has to be *timely*. It is well known that it is highly crucial to launch a new product in time which means neither too late nor too early. This timeliness of an innovation became an important driver of the cooperation and synergy realization between Phonak and Unitron.

It can be summarized that the occurrence and profitability of technology-based value creation depend on the existence of a technology-based value creation potential. This is determined by the characteristics of the resource bases and the opportunities to combine the resources to create value. The resource bases need to be attractive, well-mastered and, to a certain extent, combinable. The technology-based value-creation opportunities have to be congruent, related, feasible with the available resources and timely.

Besides these technology-related characteristics, further characteristics of the companies are required to achieve value creation in general and technology-based value creation in particular.

5.3.2 Acquisition capability

Another initial condition impacting on technology-based value creation is the acquisition capability of the acquirer. In terms of acquisition capability, the literature is mostly concerned with the acquisition experience of the company. However, most of the investigations have not found any significant correlation between acquisition experience and technological success.[26] Within this research four other critical, though related, aspects of the acquisition capability were identified:

- The formalized M&A and technology and innovation management
- The availability of resources
- The absorptive capacity
- The willingness to change.

Within this research it became obvious that a certain knowledge on the part of the M&A management, such as a formalized M&A team and process, or the internalization of these competencies via consultants, was conducive to the appropriate management of the acquisition processes and thus the realization of the value creation. Furthermore, an *established technology and innovation management*, for example managed by a CTO, is also a strong prerequisite for achieving technology-based value creation.[27] An understanding of the potential value-creation opportunities derived from combining technologies and a facilitator which can foster boundary-spanning cooperation are very helpful in the course of the acquisition and integration process. For example, the CTO from Fillpack initiated joint development projects and knowledge transfer opportunities and the senior engineer of Phonak supported the technology due diligence. Furthermore, it was observed that more innovative companies are better at achieving technology-based value creation than firms that are not innovative (Hall, 1988).

Furthermore, it was observed that the acquisition integration is highly *resource demanding*, related to not only financial resources but also skills, management attention and time. Thus an acquisition requires the availability and assignment of highly skilled personnel, sufficient financial resources and management attention. Furthermore, awareness has to be raised that integration is not over after eight weeks or six months after closing, as indicated by many M&A specialists. The integration effort lasts much longer and can be as much as up to two to five years depending on the size of the acquisition. This long-term commitment has also to be related to long-time financial investment in the integration. Without these commitments, the integration and thus value creation will not succeed. This became apparent in almost all acquisitions. For example, whereas the CEO and head of engineering from Starrag immediately moved to Heckert, at other acquisitions there was a lack of leadership personnel and integration financing and time.

The third aspect of the acquisition capability which in particular refers to technology-based value creation is the *absorptive capacity* of the acquirer.[28] As was seen within Hilti's acquisition of Ammann, it is very helpful to build up internal competencies within a technological field before acquiring a target in that field. Without internal knowledge of certain subjects it is difficult to identify, analyse, assess and especially to assimilate the target's technologies.[29] This concept, initially introduced by Cohen and Levinthal (1990), is referred to as 'absorptive capacity'. For example, Fillpack did not have much experience in transporting bottles and assessed the bottle-handling of the Aseptofill machine as profound. This incorrect estimation occurred because of Fillpack's lack of absorptive capacity. Furthermore, a certain amount of common background knowledge as was the case within the Plasma-Therm acquisitions, where Unaxis had internal competencies within PECVD technologies, is conducive to cooperation and communication. Thus it is recommended that a new technology should be acquired only if a certain amount of internal competence within this field exists to assess, assimilate and integrate specific technology.

The most important aspect, however, which has so far attained only very limited attention,[30] is the *willingness to change* on the part of both companies and particularly that of the acquirer. It became apparent that an acquisition works especially well and is conducive to technology-based value creation if it is understood as the joint creation and emergence of one new company and as an opportunity to change for the acquirer. This joint merging into something completely new provides the opportunity for the acquirer to transform into a new, more competitive and inherently balanced larger company and implies an equal relative standing between target and acquirer which fosters trust building and thus technology-based value creation. For example, Starrag saw the Heckert acquisition as its opportunity to change and described it as a win-win situation for both companies. In comparison, even though Unaxis transferred its business unit headquarter to St Petersburg, Florida, the head of the business stayed in Switzerland and the acquisition target lacked leadership. Similarly Phonak now regrets that it missed the opportunity to change accordingly along with its acquired Unitron company.

It can be concluded that an acquirer should ensure some initial M&A competence and an established technology and innovation management, sufficient available and assigned resources such as skills, management attention, financials and slack time, absorptive capacity and especially its willingness to change before initiating an acquisition.

5.3.3 Strategic fit

The third initial condition required for general value creation in corporate acquisitions is the strategic fit between the acquirer and the target. An acquisition enables the acquirer to fully obtain and own competencies in a short period of time. This exclusive ownership of knowledge can otherwise be achieved only via internal development which is mostly pursued for only highly strategic and core resources. The rapid access to competence ownership via acquisitions does not come without costs. On the one hand an acquisition is a huge financial investment and on the other hand the internalization of the acquired resources, and thus the effort to integrate the resources and to adapt the corporate contexts, often pose an underestimated obligation. Thus gaining exclusive access to and achieving internalization of competencies is highly resource-demanding no matter which strategic path internal R&D or an acquisition is chosen. Thus the acquisition's advantage of getting quick and exclusive access to competencies has to be paid for by the contextual integration effort. However, if a company does not aim for the internalization of competencies and tries to circumvent the integration efforts, then competencies do not seem strategically important enough to be accessible throughout the company. Additionally, such non-strategic competencies do not require exclusive ownership. Thus access to non-strategic competencies should not be conducted via acquisitions as these ensure exclusive ownership which is not needed. Other forms of transactions such as joint ventures, strategic alliances and the like are much more suitable for the acquisition of non-strategic resources (see Figure 5.14).

Thus it can be concluded that acquisitions should be viewed as the internalization of strategically important and fitting resources. This will ensure that integration projects and value-creation opportunities contribute to the competitive advantage of the acquirer and are seen as important projects rather than a disturbance of daily business. Additionally, the strategic importance of resources will increase management's awareness of the upcoming integration efforts and required commitment.

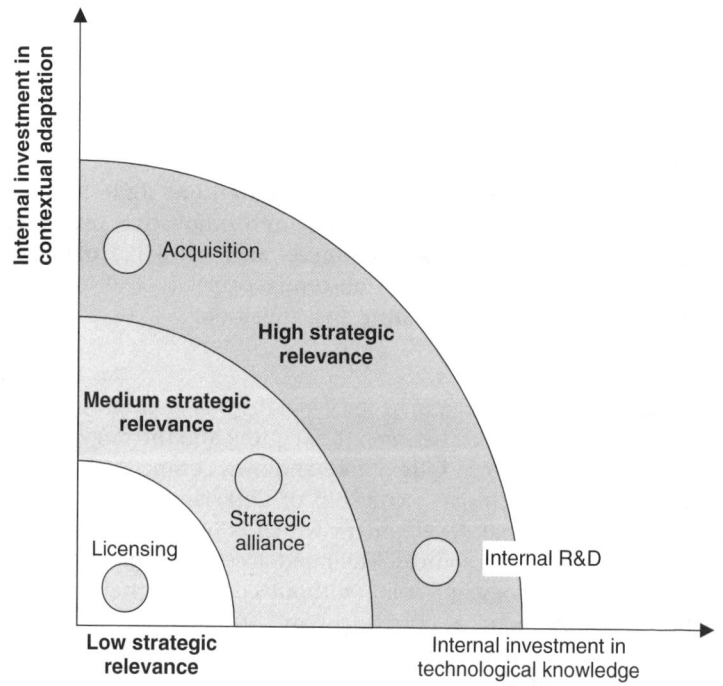

Figure 5.14 Strategic relevance of acquisitions and other transaction forms

5.3.4 Contextual fit

The fourth and one of the most important, though often neglected, initial conditions is the contextual fit between the two companies. Concepts such as a cultural or organizational fit have already been addressed in the literature (see Appendix A), but are seldom associated with technology-based value creation. Within the cases three different though related types of corporate contexts and their impact on technology-based value creation were investigated:

- Corporate culture
- Organizational characteristics
- Dominant business logic.

Within this book *corporate culture*[31] is characterized by leadership style, openness, time perceptions and the like. The *organizational characteristics* can be described as company attributes, such as level of decentralization, level of autonomy, level of formalization and the like. The corporate context of a *dominant business logic* indicates which core values and drivers are essential for the company to sustain within its business. Attributes of a dominant business logic are marketing strategies, innovation rate and business drivers.[32]

Quite interestingly there are two different findings which apply to all three contextual characteristics and were identified as crucial for technology-based value creation. The first relates to the *characteristics of the contexts* of target and acquirer and the latter to their *level of similarity* (see Figure 5.15).

It was observed that corporate contexts which have *characteristics* that foster the innovativeness of a stand-alone company are also conducive to technology-based value creation for the merged companies.[33] Thus technology-based value creation of merged companies is fostered by the innovation-promoting contexts of the individual companies, mostly because an awareness and importance of being innovative is inherent within these companies. Thus a company cannot acquire innovativeness if it does not have a context conducive to innovativeness. Within all companies in the case studies innovativeness was a major concern for competitiveness and thus all partners were aware that technological synergies will foster company success. Furthermore, cooperation between companies was facilitated if both earlier on were used to an established innovation process, R&D-related organizational structures or an open culture.

$$FIT_{CONTEXT} = f(Characteristics) + f(Similarities)$$

Figure 5.15 Contextual fit dependent on contextual characteristics and similarities

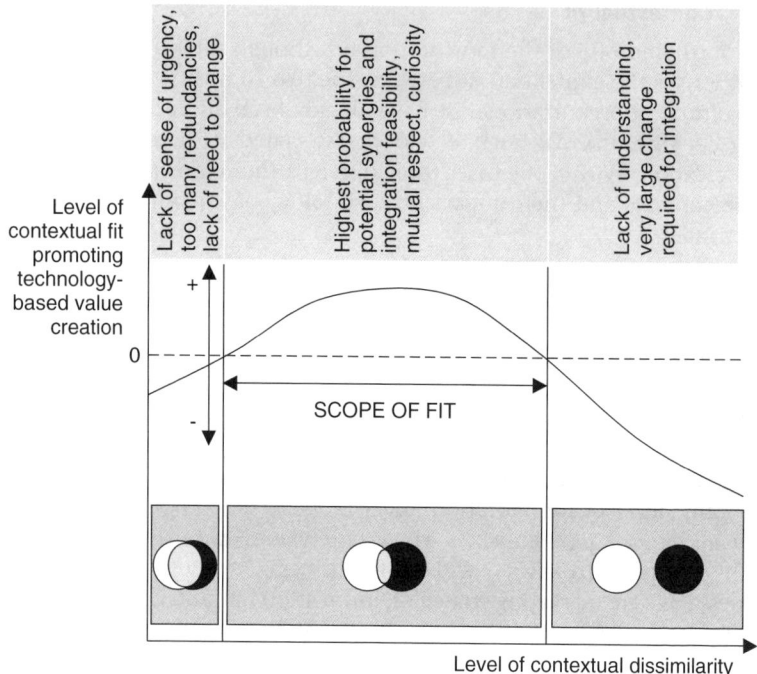

Figure 5.16 Impact of contextual similarities on technological synergies

The second finding relates to the *similarity of the corporate contexts* (see Figure 5.16). On the one hand it was found that highly dissimilar corporate contexts such as different organizational structures, corporate cultures or a dominant business logic hindered technology-based value creation. The differences, which occurred especially at the acquisition of relatively small companies, caused mistrust, very high integration efforts and a lack of mutual understanding. As already mentioned, remarks such as 'we did not really understand each other'[34] or Ammann LT's owner's misunderstanding of the structured innovation processes at Hilti characterized the cases and hindered technology-based value creation. Also differences within the organizational structures, such as Starrag's difficulties with Heckert's employment and salary system, resulted in problems with the unions. This observation is also strongly supported by other literature.[35]

On the other hand, quite interestingly, within the case studies it was also found that highly similar organizational contexts were not conducive to technology-based value creation. This, for example, was the case at the Plasma-Therm acquisition. Unaxis praised the perfect marriage and high organizational and cultural similarities with their target and thus implied the impression that the integration of the target would be comparatively easy to achieve. The deceiving perfect match resulted in a lack of sense of urgency and management commitment. Furthermore, great similarities in dominant business logic imply similar strategic directions and thus often indicate a high overlap in activities being interpreted as, and being sources of, personnel reduction which is also not conducive to successful cooperation and joint innovation projects. It is only within the organizational characteristics where a strong similarity may not cause the wrong implications. Thus a moderate to close similarity which leaves room for differences is most conducive to technology-based value creation (see Figure 5.16). The differences raise the awareness of the upcoming challenges and foster curiosity about something new rather then rejection of something substitutive or unknown.

Thus it can be concluded that technology-based value creation is facilitated if both companies already have an innovation-fostering context such as an innovation-promoting R&D structure and culture or a business logic which incorporates technology and innovation as crucial factors for competitiveness and if these contexts show a certain extent of similarity, neither too much nor too little.

From this statement it can be inferred that companies which are themselves not innovative cannot expect to acquire an innovative firm and thus easily achieve technology-based value creation; rather the opposite might occur. A firm which is not innovative can destroy the innovation-fostering context especially of a smaller partner and the acquisition may result in a sharp decrease in technology-based value creation for both companies.

As seen there is a variety of initial conditions or company characteristics which are conducive to achieving technological synergies in the course of the acquisition integration. Summarizing for the future, partner companies should have an inherent technology-based value creation potential, the acquirer should be equipped with an acquisition capability, the strategic fit should be ensured and the corporate contexts of the two companies should have an innovation-promoting character and be quite similar. Obviously these idealistic initial conditions will rarely all be fulfilled before an acquisition. However, as long as the acquirer is aware of these aspects, impact on the likelihood of successfully realizing technology-based value creation, it knows about the potential risks and can act accordingly if required.

5.4 Strategy Processes – The Appropriate Acquisition and Integration Processes

The third and most important factor impacting on technology-based value creation is the design of the acquisition and integration process. As is well known from the literature, strategy processes can highly influence the technology-based value creation achieved from an acquisition.[36] Within this chapter the focus lies on the specific criteria identified within the case studies and derived from the literature which are relevant to consider within the strategy processes to achieve technological synergies. A detailed elaboration of concepts supporting the strategy processes will be provided as a solution concept within the following chapter. The discussion of impact of strategy processes on technology-based value creation will be separated into two different phases: within the first section the impact of the process design and the main tasks of the acquisition process on technological synergies will be elaborated. Within the second section the impact of the integration approach and process on the successful realization of innovation and resource deployment opportunities will be shown.

5.4.1 The impact of the acquisition process on technology-based value creation

As implied by several researchers[37] and as observed within the case studies the design and the specific tasks fulfilled during the acquisition process have a tremendous impact on the likelihood of achieving technological synergies after the acquisition. Thus in the following text the appropriate process design and structure as well as the required acquisition tasks will be identified.

5.4.1.1 The acquisition process design

Today state-of-the-art acquisition processes are understood as highly strategic processes with non-continuous occurrence, highly influenced by external factors and partners and characterized by their limited time horizon, large information asymmetry and required secrecy. Due to these specialities it is often overlooked that the acquisition process be notwithstanding a strategy formulation process and thus needs to adhere to certain general design criteria. This oversight often results in highly secretive and financially dominated processes managed by a corporate staff team that hardly considers technology-related aspects. Thus the acquisition process should be designed according to the following criteria, derived from the case studies. In order to ensure the appropriate consideration of technological aspects, the acquisition process should be:

- Integrative
- Holistic
- Systemic.

An *integrative* acquisition process means that an acquisition is subject to not only general business strategy making, but also the marketing, HR, finance and especially technology and innovation strategy. Thus during the acquisition process all relevant areas of a company should be integrated into the strategy formulation process. Negligence of the technology dimension might lead to underestimated risks or a misunderstanding especially of long-term value creation potential, as was the case in the Aseptofill acquisition. Thus an integrative acquisition process considers all aspects of the two companies relevant for value creation. This line of argument might be opposed due to the required secrecy of the acquisition process. Indeed all measures have to be taken to ensure the privacy of the acquisition process, however without extensively compromising the quality and soundness of the strategy formulation–better losing a deal than risking a disaster.

A *holistic* acquisition process goes one step further. It not only considers the individual functions and areas but also consolidates them into one holistic picture of the acquisition. Thus highly complex interrelations, value-creation opportunities and risks can be balanced and compared. A holistic acquisition process thus ensures that technology aspects and the potential value creation and risks from combining resource bases are viewed as one integral element of the overall acquisition rational. For example, it was observed that often technological risks were identified within the technology due diligence in an acquisition but, these were not integrated within the overall company strategy and falsely were estimated as insignificant. Even though Phonak's head of engineering and senior engineer were aware of the impending risks to base a product on the very new outsourced DSP factory platform and reported this to the board, the board did not take sufficient notice of this threat and declared that the technological risks could be mastered. Finally the developed product was delayed in its launch by six months and never really became a success. This example shows that lack of a holistic integration of technology aspects into the overall acquisition strategy making can have poor results.

Additionally, the acquisition process needs to be *systemic*.[38] It was observed that an established acquisition process supported by an M&A team or by consultants was conducive to acquisition success and technology-based value creation. It ensures the diligent fulfilment of the acquisition tasks and the integrative and holistic procedure of the acquisition process in a timely, but not too quick, manner. It determines who should participate and when during the acquisition process in a systematic way and thus facilitates the whole strategy-building process – an example was the Plasma-Therm acquisition process.

Additionally, a systemic, holistic and integrative acquisition process facilitates clear transition to integration and thus strategy implementation processes, which is essential to realizing the developed strategy and planned value creation.

5.4.1.2 The acquisition process structure

In addition, this systemic, holistic and integrative acquisition process needs to be supported by the appropriate acquisition structure which actively promotes the identification, assessment and realization of technological synergies. For example, Foster and Kantrow (1988: 51) state that one should 'not underestimate the power of an effective chief technical officer [to achieve technology-based value creation in corporate acquisitions] when he is allowed to function as a corporate strategist'. Furthermore, the integration of gatekeepers[39] to support the integration process[40] or other boundary-spanning technology-related functions are proposed.[41] Another structural issue is the assignment of a team of managers specifically in charge of overseeing the transfer of knowledge and the achievement of technology-based value creation (Kozin & Young, 1994).[42] Furthermore, the development of information systems that link R&D with key stakeholders across the business units or of incentive systems that foster R&D cooperation is recommended (Hitt, M., Hoskisson, R. E., Ireland, R. D., et al., 1991b).

5.4.1.3 The acquisition process tasks

Besides the appropriate acquisition process design and underlying structure, the tasks which are fulfilled during the acquisition also have an important impact on the likelihood of achieving technological synergies. These tasks comprise:

- *strategy development*
- *assessment of the technology-based value-creation opportunity*
- *integration planning.*

Diligent, prospective *strategy formulation* is a core task of the strategy formulation process. This task also needs to be mastered in an integrative, holistic and systemic manner and should integrate the technological aspects as well as identify and assess the potential value-creation opportunities based on realistic scenarios. As Hitt *et al.* (1991b: 32) and others[43] state, 'executives … can follow an acquisition strategy and be innovative, but only with careful planning and execution'. Chaudhuri and Tabrizi (1999: 195) make this statement: 'successful acquirers, by contrast, systematically determine and outline their capability needs'. Furthermore, Kozin and Young (1994) state that potential opportunities for future offerings based on the capabilities of the companies have to be analysed. Similarly James and his colleagues (1998: 569) recommend the acquirer: 'think beyond the deal to consider the implications of the acquisition for technology in both the acquirer and the acquired business and the post-acquisition approach that can best realize value from the post-acquisition combination of their technological assets'.

The second task, the diligent, prospective assessment of the technologies also referred to as *technology due diligence*[44] shall overcome the information asymmetry between target and acquirer, ensure the absence of technology-related risks, validate the potential for technology-based value creation and support the acquisition price finding. Without a diligent technology due diligence, upcoming technology-related risks might be underestimated (as was the case in the Aseptofill acquisition), the value creation potential misunderstood (as happened in the Plasma-Therm and Unitron acquisition) and the acquisition price falsely calculated.[45] Other authors[46] argue that the maturity of the technology in question highly effects successful value creation, and thus needs to be investigated during the due diligence. Additionally, an analysis of the candidate's existing products and capabilities are an indication of the technological capabilities (Chaudhuri & Tabrizi, 1999). Furthermore, the strengths of the underlying IP assets, infrastructure and R&D capabilities and key people are considered to highly impact on the potential for technology-based value creation.[47] Other factors to be considered are external factors such as market trends[48] or technology trends. Thus the integration of marketing experts and knowledge seems appropriate at the technology due diligence investigation.[49] The integration of immature technologies, for example, especially if the acquirer is not aware of this flaw, can result in poor R&D performance and innovativeness.[50] A few authors[51] have specifically addressed this aspect of the technology due diligence and developed concepts to conduct a technology due diligence which aims to improve the potential for technology-based value creation and thus reduce the risks of inappropriate investments in companies. The approaches of the authors are highly varied in focus, applicability and detail. Most of the concepts investigate attractiveness of the current technologies and their relation to past success. The more future-oriented perspective within the technology due diligence, the investigation of whether the technologies can be the basis for future technology-based value creation, is hardly addressed.

The third task is diligent and detailed *integration planning*. A profound and holistic integration planning was often a key success factor for acquisition success and technology-based value creation. It is needed to ensure the link between strategy formulation and implementation and can thus determine success or failure. The negative results of a lack of integration planning became apparent in almost all the case studies.

Thus it can be concluded that an integrative, holistic and systemic acquisition process which includes the diligent fulfilment of strategy building, due diligence and integration planning is conducive to technology-based value creation after the acquisition. Concrete examples of how to fulfil these tasks will be provided within the next chapter on technology-based strategic acquisition and integration management.

5.4.2 The impact of the integration process on technology-based value creation

As mentioned, observed and confirmed by various researchers, the integration and thus strategy implementation process is the main determinant of acquisition success and failure and of the successful realization of technology-based value creation. Various authors have tried to explain which integration approach should be chosen and how the integration process shall be designed to achieve general value creation; however, only a few authors have so far attempted to explain the impact of the integration approach and process on technology-based value creation and these are highly inconclusive and even partly contradictory.

A few researchers state very clearly which integration approach is favourable but some of them are not in agreement. For example, researchers such Jemison (1988) or Chaudhuri and Tabrizi (1999) plead for a low-integration approach, if technology-based value creation is pursued. They argue that innovation and the leveraging of technologies requires learning, mutual respect, trust[52] and thus needs time. Whereas others[53] have observed that only a tight integration approach leads to the expected results. They claim that technology-based value creation requires a high level of interaction, communication and cooperation.[54] For example, Capron and Mitchell (1998) have observed that interaction and bilateral redeployment of resources improve the R&D capabilities of the firms.[55] Chakrabarti and his colleagues (1987; 1994) modify this general statement on the tight integration approach by saying: 'A high level of integration with the corporation was seen to have high positive relation with performance ... Organizational integration without an excessive increase in formalization was found to be key to enhancing the performance of the acquired division' (Chakrabarti & Souder, 1987: 113; Blex & Marchal, 1990).

Birkinshaw (1999) and, similarly, Puranam and his colleagues (2003) distinguish between two different ways of achieving technology-based value creation which resemble a trade-off between human and task integration.[56] Also Ranft and Lord (2002: 430) describe the integration antagonism by stating that 'greater autonomy during integration facilitates the preservation of an acquired firm's tacit and/or socially complex knowledge ... [however] ... inhibits the transfer of the acquired firm's technologies and capabilities that are based on tacit and or socially complex knowledge'.

Birkinshaw (1999) distinguishes between a high and a low road to integration. The low road describes a fast level of task integration, risking demotivation of people, cultural clashes and experts leaving, however achieving operational synergies quite soon after the acquisition. The high road takes more care about people's behaviour. Integration at the high road level pursues a quite distant approach resulting in late operational synergies and often in a second wave integration. Regarding technology-based value creation, Birkinshaw (1999: 38) concludes that: 'taking the high road of slow and cautious post-acquisition integration is the best choice when the acquired company is knowledge-intensive'.

Puranam and his colleagues (2003) have investigated the impact of different integration approaches on innovativeness. They have come to the conclusion that a low-integration approach results in fast initial innovation outputs; however, subsequent innovativeness is limited. At a high level of integration the time-to-market for the first new product takes longer, but innovativeness remains stable. Thus in contrast to Birkinshaw's (1999) findings it can be concluded that a tight integration approach and thus the low road would be favourable with respect to technology-based value creation.

Birkinshaw's model (1999) also shows the relation between integration approach and speed. Thus the integration approach is related to integration behaviour. Whereas some authors[57] are in favour of a low speed of integration and in particular technology integration, providing slack to enable knowledge preservation and learning, others[58] have observed that a high speed is more appropriate. Foster and Kantrow (1988) make a more relative statement by recommending an appropriate level of management pressure.

Another aspect unresolved and intensively discussed in integration planning is the integration structure. This aspect is mainly concerned with retention of the management of the acquired company and the leadership structure during integration. Ranft and Lord (2002), for example, state that the proportion of managers from the acquirer within the integration management reduces the retention rate and is curvilinear to communications. Best practice reports and researchers[59] conclude that the managers of the acquired company may need to change, whereas others[60] recommend keeping the leaders of the purchased company.[61]

The only areas of agreement within the integration planning comprise the importance of a shared technology vision and strategy,[62] the retention of key engineers[63] and a high level of communication, the use of symbolic actions[64] and a fair integration behaviour.[65] Also the positive impact of story-telling versus the negative one of rumours is discussed.[66] Furthermore, the early decision upon the integration approach and management and thus the importance of early and appropriate integration planning is agreed upon.

In order to bring some clarification to the above observations and to discuss the respective findings from the case-studies, a model (see Figure 5.17) is provided to guide elaboration and thus reduce complexity. This model was observed in all case studies, matches the findings from the literature and was discussed with various practitioners and researchers and confirmed as representing reality. It allows the explanation of the impact of the acquisition type on the integration processes and of the occurrence of innovativeness and efficient resource deployment in the course of the integration progress. Additionally, the model provides an overview of the overall developments within the whole company over the time of the integration progress.

On the x-axis of the model the integration progress is shown. The integration progress and not the integration time is deliberately provided as these two factors are often falsely interrelated. Integration progress represents the level of achieving corporate coherence[67] between the acquirer and the target. On the y-axis the cumulated internal frictional resistance is drawn. This is an indication of the internal inefficiencies within the merged companies. It could be measured by redundancies, inefficiencies in process and information flows, internal human resistance and the like.

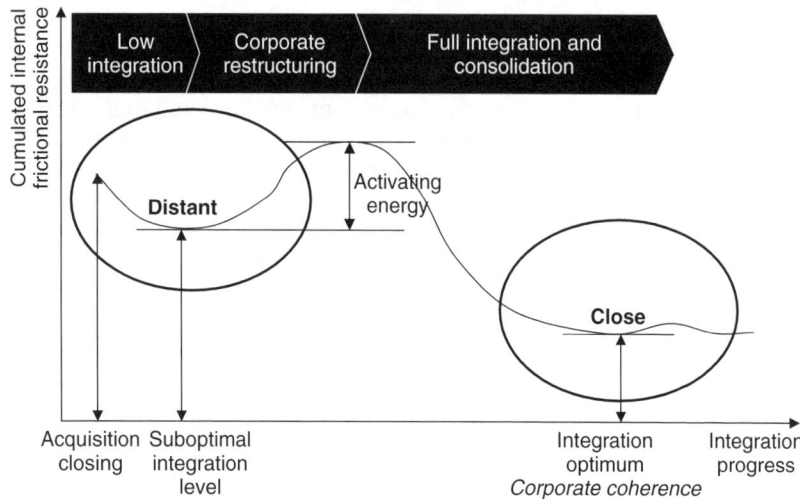

Figure 5.17 Model of the acquisition integration process

This model of the integration progress shows that all acquisition integrations undergo an evolution or development rather than occurring simply as an event. Each integration stops at a different phase of this evolution. This development can be separated into three phases: the *low integration* phase; the *corporate restructuring* phase; and the *consolidation or full integration* phase finally achieving dynamic corporate coherence. Based on this model it will be explained in which phase of the integration evolution which technology-based value creation occurs and what it is dependent upon. From this can be inferred which aspects are especially important to consider in order to achieve technology-based value creation in the course of the integration progress.

5.4.2.1 The low integration phase

The first phase, low integration, is normally referred to as the typical integration phase. It starts with the closing of the acquisition and is mainly concerned with the realization of initial synergies, such as synergies within the procurement or sales area. Furthermore, the typical technology-based quick gains, such as cross-selling opportunities, are achieved. Additionally, the supportive integration, such as integration of the HR and IT systems, the financial and legal integration and the like, is conducted. Organizational integration is mostly conducted to a very limited extent. The phase ends with the typical low-integration approach. The cumulative internal frictional resistance is reduced only a little as low organizational integration results in inefficiencies in the organization and redundancies prevail. Behaviour during the initial phase should be focused on retention and communication. The main focus lies on sustaining value and keeping people as well as persuading them of the vision and strategy of the new joint company. Very often during this phase the old leadership personnel of the target company is retained to master the first adaptations and to motivate the employees. The leadership style though should be very strict and dominant to reduce chaos and to achieve the required workload during this intense integration time. Thus it is recommended that a new leadership team should be introduced to the target right from the beginning of the integration. This first low-integration phase lasts very differently at the various acquisitions. Many companies today pursue this low-integration approach as a permanent status and retain the acquired company at a distance, even though the potential for value creation is limited to quick gains. Occasionally, for example as occurred with IBM's acquisition of Lotus Notes, this at-arm's-length integrated and quite independent target loses its innovativeness after some time as it cannot work for itself and many key people leave after several years following the acquisitions. Then the performance of this independent daughter company decreases and additionally the mother company increasingly feels the urge to reap further synergies. This often results in a second-wave integration, mostly around five years after the initial integration.[68] The large problem with this second-wave integration is the lack of sense of urgency to change among the target's personnel and its resistance to this very late integration effort.[69] Thus the initial precautionary having low integration, often justified by arguments such as: 'If we fully integrate them, then they lose their innovativeness or their key people will leave', then turns into even more competence destruction several years after the acquisition. Thus it might make sense only in acquisitions that aim solely for value creation from quick gains with a limited time horizon, which is sometimes the case at venturing acquisitions, to keep them in the low-integration phase. All other acquisitions which also aim for long-term technology-based value creation should further evolve within the integration development.

5.4.2.2 The corporate restructuring phase

The second phase is the corporate restructuring phase. As the low-integration phase results in a suboptimal level of integration of the two companies, within the corporate restructuring phase additional effort has to be taken to change and restructure both companies as one new joint enterprise with a new configuration. This corporate restructuring requires both acquirer and target to change their corporate contexts in order to build a new and better enterprise and to use the joint resources and competencies. This restructuring, however, requires additional effort and temporarily increases the accumulated internal frictional resistance but without this additional effort (in chemistry the analogy of 'activating energy' would be appropriate) the corporate coherence of one new emergent company cannot be achieved (see Figure 5.17). In the course of this corporate restructuring phase the wheel of technology-based value creation starts spinning. The restructuring enables the redeployment and substitution of resources such as the co-location of engineering teams or the transfer of manufacturing activities. Furthermore, initial innovation activities can be initialized. Besides this the corporate restructuring phase is concerned with the reorganization of the corporate processes and structures and finally enables internal coherence within the new company and its fit with external developments. Additionally, the integration has to focus on change and retention management. During this highly complex and resource-demanding time, the integration process has to have top management's full attention. Furthermore, the key personnel and particularly 'change agents' have to be involved in achieving organizational change, and to communicate a new joint vision and strategy as a guideline for company transformation. The integration has to have a momentum and should be supported through efficient communication, story-telling and the publication of the successes. Management style has to be decisive, fair and clear. This integration behaviour during the corporate restructuring phase can be understood as a 'catalyst', again referring to the chemistry analogy, to lower the required activating energy. This process has to be mastered very diligently and actively and is best done by a new leadership individual, especially as there is the risk that the integration could revert to the suboptimal integration level. This was, for example, the case at the Plasma-Therm acquisition. The transfer of the business unit headquarters was a good start for a corporate restructuring, but its poor execution resulted again in a very low level of integration and limited potential for technology-based value creation – in that case the development of the Unified Platform.

The duration of the corporate restructuring phase varies greatly. It starts partially immediately after the closing or several years after the acquisition as a second-wave integration effort and its end is determined by the commitment and management of the restructuring phase. Furthermore, the restructuring phase can occur in different company areas at different times. Whereas the full integration of sales channels often occurs quite soon after the integration, the restructuring of innovation and technology-related functions often starts much later and takes longer.

5.4.2.3 The consolidation phase

The third integration phase is the full integration phase or consolidation phase. Within this phase the focus is on not the organizational change but the slow merging and consolidating of the two companies into one new enterprise. A new joint corporate context, such as a new corporate vision, strategy and culture, slowly emerges. Hand-in-hand, fostered by interaction and confrontation during the restructuring phase, emerge trust and mutual understanding. Thus the wheel of technology-based value creation starts spinning even faster. The focus is on achieving a balanced company pursuing corporate coherence and realizing long-term technology-based value-creation opportunities.[70] Technology and product platforms occur and competence centres are formed. In the course of the consolidation phase the accumulated internal frictional resistance slowly decreases until it reaches a dynamic corporate coherence. The organizational structures are either fully integrated or contain boundary-spanning structures. The integration behaviour is focused less on dominant leadership and rapid pace and more on providing slack to achieve mutual understanding and technology-based value creation. Thus, again using the chemistry analogy, it acts as a 'moderator'. Only then, mostly several years after the acquisition, does the company reach a coherent state, whereas mostly this stage is never achieved due to other strategic undertakings. However, it is important to recognize that an acquisition is not finalized eight weeks or half a year after closing; its impact lasts long after the contract has been signed.

Thus it was observed that the design of the integration phase and associated activities have a crucial impact on technology-based value creation. Whereas in the first low-integration phase only quick gains can be achieved, a second phase of corporate restructuring is required to achieve a joint company configuration which enables long-term technology-based value creation. Within each phase different aspects have to be addressed and activities have to be mastered. Furthermore, the integration behaviour, such as leadership style, integration pace, communication and the like, plays a different role within the different integration phases. It can be concluded that acquisitions which pursue short- and long-term technology-based value creation need to undergo all three phases of the integration evolution to reap full value creation potential.

It can be summarized that the acquisition and the integration processes have a significant impact on technology-based value creation. Whereas an integrative, holistic and systemic acquisition process and the diligent fulfilment of the associated tasks ensure the appropriate consideration of technology-based value creation potentials, the management of the integration evolution finally determines its realization over time and subsequently the generation of revenue.

5.5 External Developments

The fourth main aspect impacting on technology-based value creation besides acquisition type, initial conditions and strategy processes, is external developments. A variety of different impacts of market trends and other external developments on technology-based value creation were observed. A slow market development provided slack especially to achieve resource deployment, learning and trust building between the two companies, as was the case at the Plasma-Therm acquisition where, after the boom in the semiconductor industry, the downturn time was used to develop the product platform. A faster-than-expected market development gave rise to the urge to reap technological synergies, as was the case at the Unitron acquisition; however, this also posed the threat of asking too much from R&D personnel. For example, during the first year of the Plasma-Therm acquisition activities were entirely focused on serving customers' needs and hardly any technological synergies were achieved. This was justified by the booming market. Thus it can be inferred that market developments should be used as they come. In highly dynamic times the focus should be on sustaining and increasing the existing value while reaping easily achievable technological synergies: previously referred to as quick gains. Accordingly slower market developments should be used to redeploy the resource base and to reap resource deployment synergies.

5.6 Conclusion from the New Understanding

This model of understanding provided the answer to the first research question on which aspects are relevant for achieving technology-based value creation in corporate acquisitions in innovation-driven industries. Details were given of how technology-based value creation occurs, the acquisition types in which it becomes relevant and in which way it needs to be addressed. Furthermore, the initial conditions, an inherent technology-based value creation potential, an acquisition capability, a strategic fit and a contextual fit between target and acquirer that are required for technology-based value creation were described. Additionally, the correct design and mastery of acquisition and integration processes for promotion of resource redeployment and innovation achievment was described. Last but not least, the impact of external development on the occurrence of technological synergies was discussed. These findings were based on cases from practice and enhanced by notions from the literature. In the next chapter a concept for strategic acquisition and integration management, taking into consideration the above findings, will be developed.

6
Technology-based Strategic Acquisition and Integration Management

The previous chapter on the new understanding of technology-based value creation in corporate acquisitions outlined the criteria which are required to achieve successful resource deployment and innovativeness from acquisitions. However, practical concepts for applying the criteria identified in the course of the strategic acquisition and integration management are still missing. Thus this chapter aims to close this gap and introduces a theory for technology-based strategic acquisition and integration management, thus answering the second research question guiding this book. This technology-based strategic acquisition and integration management theory aims to provide managers with concepts that support their decision-making in corporate acquisitions in innovation-driven industries. The concepts help to integrate the identified criteria required for successful technology-based value creation into acquisition and integration management. Thus they support the development of growth options which provide long-term value creation potential, the selection of potential targets which offer a joint technology-based value creation potential and the development of a technology-aware acquisition and integration strategy. This integrates future technology and innovation-related activities and helps to validate the occurrence, profitability and successful realization of technology-based value-creation opportunities in the course of the technology due diligence. Furthermore, the concepts support managers in determining and planning an integration approach which fosters the realization of long-term technological synergies. Thus the technology-based strategic acquisition and integration management theory helps companies to become more innovative after their acquisitions and supports them in efficiently deploying their capabilities.

Within a first section of this chapter the development of the concept is outlined. In a second section an overview of the concept and its integration is provided and in a third section the different phases and layers of the technology-based strategic acquisition and integration management are discussed in detail.

6.1 Developing the Concept

The technology-based strategic acquisition and integration management is designed as a strategic process aiming to *formulate and implement an acquisition and integration strategy*. Thus it consists of a strategy formulation process: the acquisition process, and a strategy-implementation process: the integration. The acquisition process is a prescriptive and rational planning process as well as a political and social process of interactions between people.[1] The integration process as strategy-implementation process is understood as a process of action, measurement and control concerned with the design of the structures, processes and behaviour of the organization influenced by the integration leadership and context.

Furthermore, technology-based strategic acquisition and integration management will be an *adaptation* of the *existing* strategic acquisition and integration management concepts found in theory and practice. Thus it will be structured within pre-transaction, transaction and post-transaction phases and will adopt several aspects which are not related to technology aspects in acquisitions.

Additionally the technology-based strategic acquisition and integration process will *fulfil certain requirements*. Thus it has to actively consider and integrate the criteria that are conducive to successful technology-based value creation from corporate acquisitions. The process will consider the impact of different acquisition types on resource deployment and innovativeness, it will assess the initial conditions to foster technology-based value creation, it will adhere to the innovation-fostering process design and consider the impact of external developments on technology-based value creation (see also Chapter 5).

Furthermore, the concept of technology-based strategic acquisition and integration management will be elaborated in *two different steps*. The first provides a *rough overview* of the process as strategy formulation and implementation concept, addresses the *holistic perspective* including several company areas and is used to outline the *integrating nature* of the concept between strategic management, acquisition and integration, and technology and innovation management. The second step is used to specifically focus on the *aspects relevant for technology-based value creation* in corporate acquisitions. It addresses each of the six process phases and the four different layers of the acquisition and integration process individually, explains structures and methodologies useful for achieving technology-based value creation and provides implications for the behaviour during the process phases. These individual elements, while they are embedded within the overall process, however are thought to be used and applied selectively and also individually. Thus the proposed elements can be chosen according to specific needs and also applied within other strategy processes. Thus the detailed concept shall be used as a modular framework of 'plug'n'play' elements. Additionally, the detailed concepts will be further explained and exemplified by best practices from industry.

6.2 Overview of the Concept

The technology-based strategic acquisition and integration management consists of a strategy formulation and implementation process split into six different phases (see Figure 6.1). These phases cover the acquisition and integration process in an integrated and holistic way. Thus the objective of the process is to develop and to implement an acquisition and integration strategy. The process mainly takes place on the strategic level; however, it is also intertwined with the normative and operational level. Furthermore, the process is managed on different layers represented by the core M&A team, the functional teams, the supportive teams and the integration team. Additionally the process bridges the gap between technology and integration management on one hand and acquisition and integration management on the other. These characteristics will be outlined in the following.

6.2.1 Phases of the technology-based acquisition and integration management

The first and initiating phase of the acquisition process is the regular strategy planning of the acquirer. Thus it is supposed that an acquisition should not be opportunistic but the result of a systematic and holistic business-strategic planning. This ensures consistency between short- and long-term business strategies and the acquisition rational. Once the idea of acquiring a company fits with the company strategy, the actual acquisition process is initiated with the task of screening for a company. Many large companies today have a strategic M&A department which is in charge of regularly searching for potential targets. If this is the case, then the first two phases of the process may run in parallel. After having identified a potential and attractive target, a core acquisition team is set up. This mainly consists of the business unit manager, a financial controller, an M&A manager, tax and legal experts, the CTO and a transition manager, which will be responsible for the integration.

Subsequently the third phase begins. The core team and its subteams are in charge of developing a rough and preliminary acquisition and integration strategy – or often referred to as a business plan based on assumptions and financial figures – and of introducing it to get an approval to proceed with the acquisition. The pre-transaction phase lasts until this approval is obtained. Then the target company is contacted, a non-disclosure agreement is signed and the acquisition and integration strategy will be detailed based on the reducing of information asymmetry. This phase is a repetition of the third phase: the acquisition and integration strategy development. Another approval will be required to enter the next phase which aims to further reduce the information asymmetry and to detail and revise the developed acquisition and integration strategy. This phase is called the due diligence. This due diligence is concerned with validating the acquisition and integration strategy, with assessing the risks and supporting the price finding. Within the further discussions the technology due diligence will be one core element.

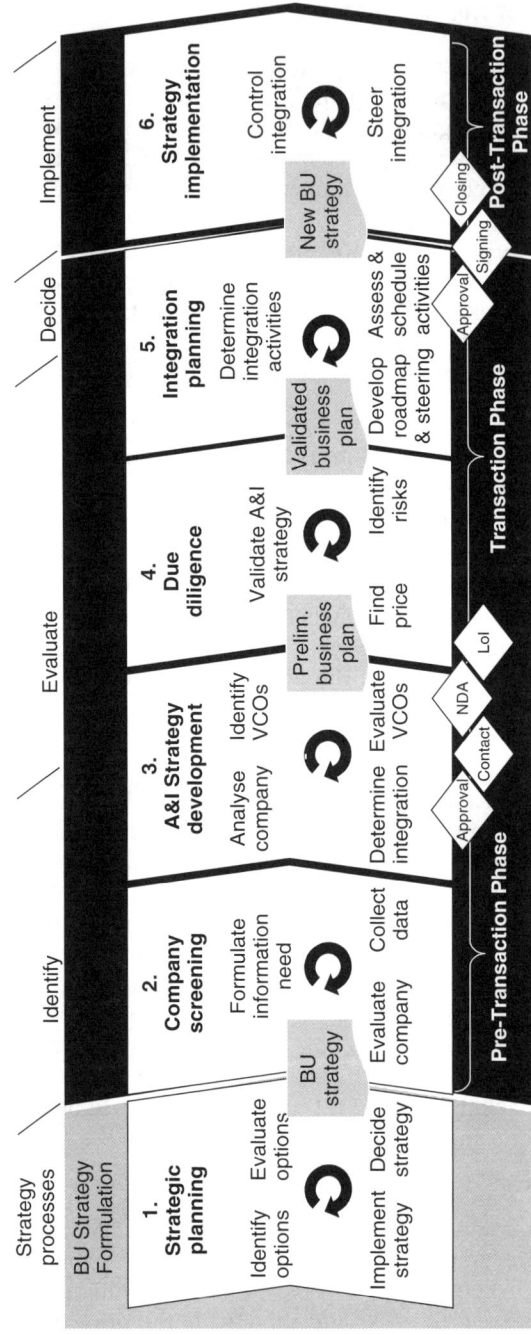

Figure 6.1 Technology-based strategic acquisition and integration management

After the acquisition, or possibly in parallel with it, the fifth phase, concerned with the integration planning, starts. After these phases the transaction and strategy formulation is finalized with the signing of the acquisition contract. Subsequently, after the closing, the integration phase, which represents the actual implementation of the new joint business strategy, is started on the operational level and steered and controlled on the strategic level.

An important characteristic of the strategic acquisition and integration process are its *parallel activities*. Thus the process phases are not as sequential as indicated but rather they overlap and periodically iterate with the reducing of information asymmetry. Another crucial aspect is the *interrelation between the acquisition and the integration process*. Thus this technology-based strategic acquisition and integration management sees strategy formulation and implementation as highly linked processes which need to be understood as two elements of one strategic acquisition. An example of the interrelatedness between acquisition and integration process can be seen within the best practice case of General Electric's acquisition and integration management (see Best Practice 1).

6.2.2 Layers of the technology-based acquisition and integration management

The strategic acquisition and integration process can also be divided into layers. Generally four different layers with different tasks and represented by different teams can be distinguished. The main layer which coordinates the other layers is the management layer represented by the core M&A team (see Figure 6.2). This project management layer is responsible for the overall development and implementation of the acquisition and integration strategy and coordinates and aggregates information from the different sub-layers. These are the functional layer, the supportive layer and the integration layer.

The functional layer, represented by the various functional subteams of the core M&A team, is in charge of the functional content of the acquisition and integration. Thus the sales and marketing team, the manufacturing team, the supply chain and procurement team as well as the R&D team are concerned with formulating their information needs, with identifying and evaluating value-creation opportunities, with the due diligence within their area and with integration considerations within their function. Furthermore, the functional teams coordinate, steer and control the operational integration within each company function. The supportive subteams are concerned with company valuation, tax and financial investigations, communicating the acquisition and integration, negotiation of the deal and the change management associated with the acquisition and integration. The teams support the core and the functional teams and thus complement the development and implementation of the acquisition and integration strategy. Last but not least, the acquisition and integration process also takes place

Revenue in 2002: USD 131.7bn, Net earning 2002: USD 15.1bn; # employees: 315,000
Businesses: Advanced Material, Finance, Consumer & Industrial; Energy, Equipment Service, Healthcare, etc

General Electric is highly experienced in doing acquisitions: 'we consummate about two acquisitions a week' (GE Capital, 2001). The acquisition and the integration process are two integral parts of the overall acquisition and integration management. GE continuously improves the acquisition capability.

Pre-acquisition	Foundation building	Rapid integration	Assimilation
- Due diligence - Negotiation and announcement - Close	- Launch - Acquisition integration workout - Strategy formulation	- Implementation - Course assessment and adjustment	- Long-term plan evaluation and adjustment - Capitalizing on success
• Begin cultural assessment • Identify business's cultural barriers to integration success • Select integration manager • Assess strengths / weaknesses of business and function leaders • Develop communication strategy	• Formally introduce integration mgr • Orient new executives to GE business rhythm and non-negotiables • Visibly involve senior mgmt • Jointly formulate integration plan including 100-day and communication plans • Provide sufficient resources & assign accountability	• Use process mapping, CAP and workout to accelerate integration • Use audit staff for process audits • Use feedback and learning to continually adapt integration plan • Initiate short-term management exchange	• Continue developing common tools, practices, processes and language • Continue longer-term management exchanges • Utilize corporate education • Use audit staff for integration audit

Source: Capital, G. (2001) *The Right Way – and Some Wrong Ways – to Make an Acquisition*, GE Corporation.

Best Practice 1 Acquisition and integration management at GE

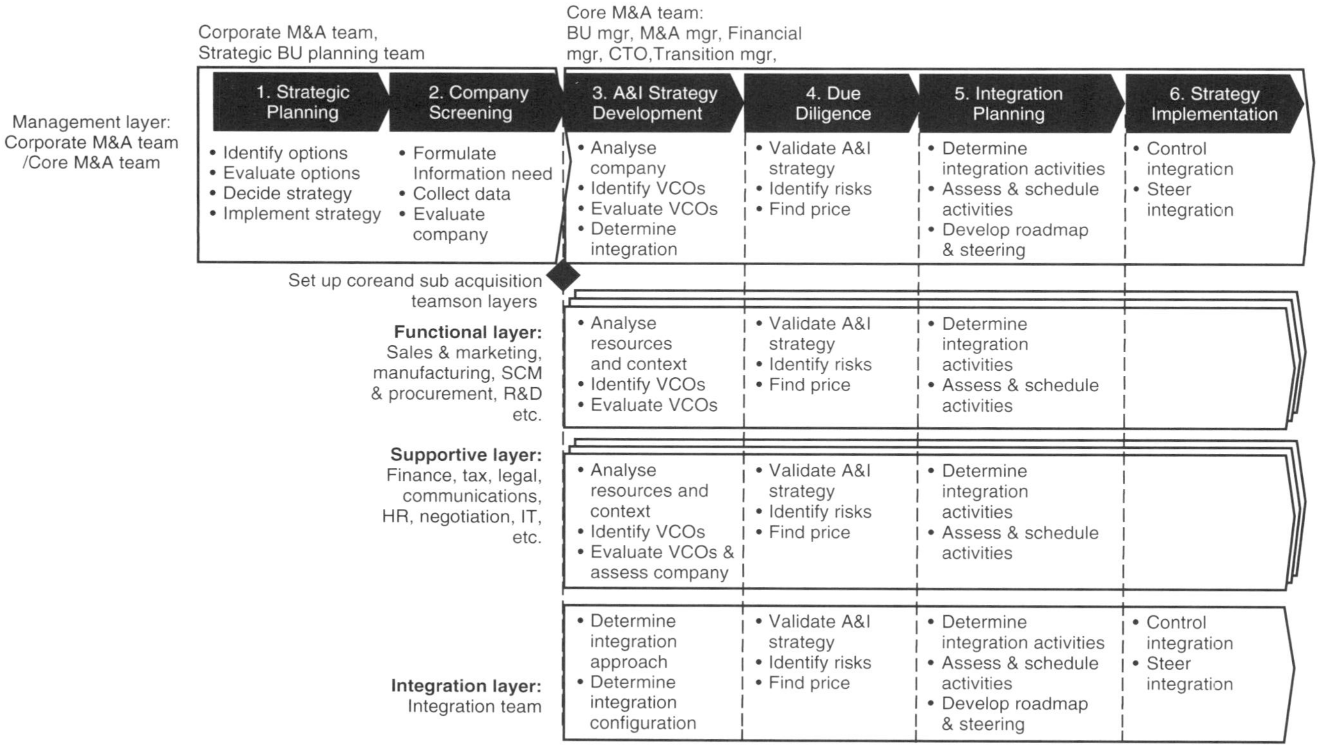

143

	1. Strategic Planning	2. Company Screening	3. A&I Strategy Development	4. Due Diligence	5. Integration Planning	6. Strategy Implementation
Management layer: Corporate M&A team /Core M&A team	• Identify options • Evaluate options • Decide strategy • Implement strategy	• Formulate information need • Collect data • Evaluate company	• Analyse company • Identify VCOs • Evaluate VCOs • Determine integration	• Validate A&I strategy • Identify risks • Find price	• Determine integration activities • Assess & schedule activities • Develop roadmap & steering	• Control integration • Steer integration
Functional layer: Sales & marketing, manufacturing, SCM & procurement, R&D etc.			• Analyse resources and context • Identify VCOs • Evaluate VCOs	• Validate A&I strategy • Identify risks • Find price	• Determine integration activities • Assess & schedule activities	
Supportive layer: Finance, tax, legal, communications, HR, negotiation, IT, etc.			• Analyse resources and context • Identify VCOs • Evaluate VCOs & assess company	• Validate A&I strategy • Identify risks • Find price	• Determine integration activities • Assess & schedule activities	
Integration layer: Integration team			• Determine integration approach • Determine integration configuration	• Validate A&I strategy • Identify risks • Find price	• Determine integration activities • Assess & schedule activities • Develop roadmap & steering	• Control integration • Steer integration

Corporate M&A team, Strategic BU planning team

Core M&A team: BU mgr, M&A mgr, Financial mgr, CTO, Transition mgr,

Set up core and sub acquisition teams on layers

Figure 6.2 Different layers of the technology-based acquisition and integration process

144 *Mastering the Acquirer's Innovation Dilemma*

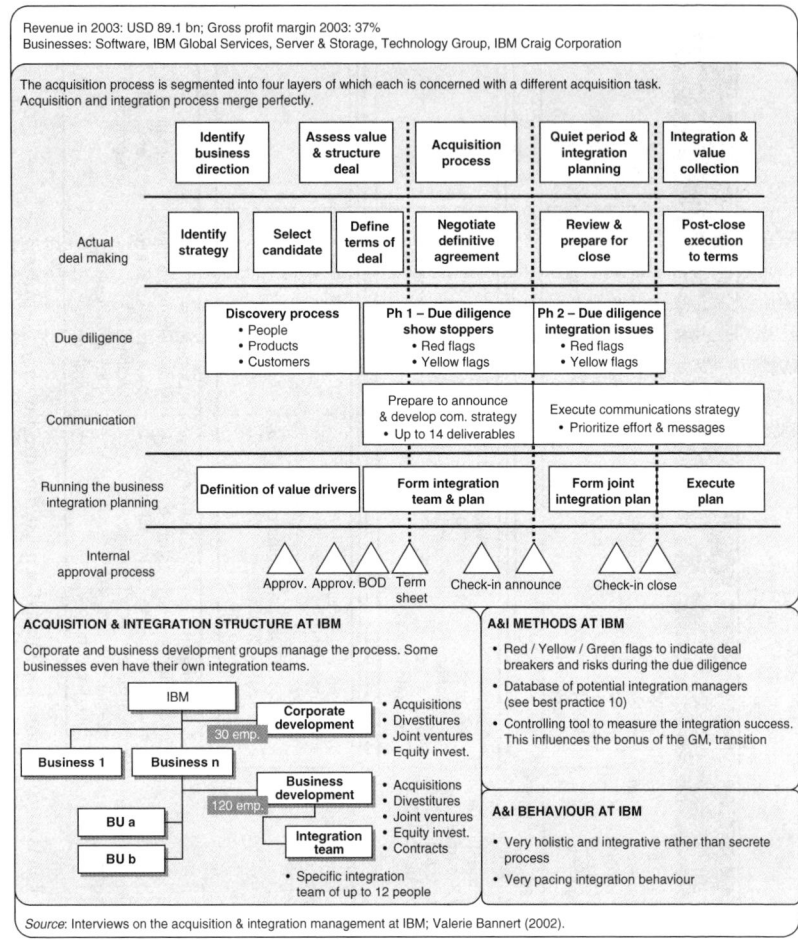

Best Practice 2 Acquisition and integration management at IBM

on an integration layer represented by the integration team headed by the transition manager. This layer ensures the continuous and consistent consideration of the integration aspects throughout the acquisition and integration process and is in charge of assessing the contextual fit, of developing an appropriate integration approach and configuration, and of planning, implementing and controlling this approach.

An example of the acquisition and integration process in its different phases and on different layers can be seen in Best Practice 2 on IBM.

6.2.3 Technology-based strategic acquisition and integration management on the normative, strategic and operational level

Whereas the main acquisition and integration process takes place on the strategic level, it is also tightly linked to the normative and operational level (see Figure 6.3). Generally the normative elements of a company, such as a long-term vision and a company policy, impact on all phases and layers of the technology-based acquisition and integration process. Thus these general guidelines influence strategy building, selection of a matching target, acquisition and integration strategy and particularly integration management. Additionally the acquisition and integration process is also linked to other strategic processes. It is initiated by the general strategic planning processes and strongly affected by other strategic initiatives of the acquirer such as reorganizations, other acquisitions or the like. Furthermore, the acquisition and integration process needs to be supported by and based on the operational processes. Thus general strategic planning is based on operational advances, company screening needs to be fed by the experience of the functional areas of the company, and acquisition and integration strategy development needs to be based on the judgment of information gained from operational line managers. Last but not least, integration mainly takes place on the operational level. Thus successful acquisitions can be fostered by consistently integrating operational functions into the whole acquisition and integration process.

6.2.4 Integrating the technology and innovation management into the AIM

Additionally, this overview of the concept of technology-based strategic acquisition and integration management is used to outline its integral character linking technology and innovation management to both strategic management and acquisition and integration management (see Figure 6.4). The individual aspects and their interrelation will be addressed within the detailed discussion of the different phases of the technology-based strategic acquisition and integration process below.

6.3 The Detailed Concept

Within this chapter each phase of the technology-based strategic acquisition and integration process will be detailed and discussed. The main objectives of each phase as well as the critical aspects to consider for achieving technology-based value creation are outlined. Furthermore, different processes, methodologies, structural and behavioural elements and implications are introduced. These elements will complement, enhance and partly substitute the existing acquisition and integration management and foster

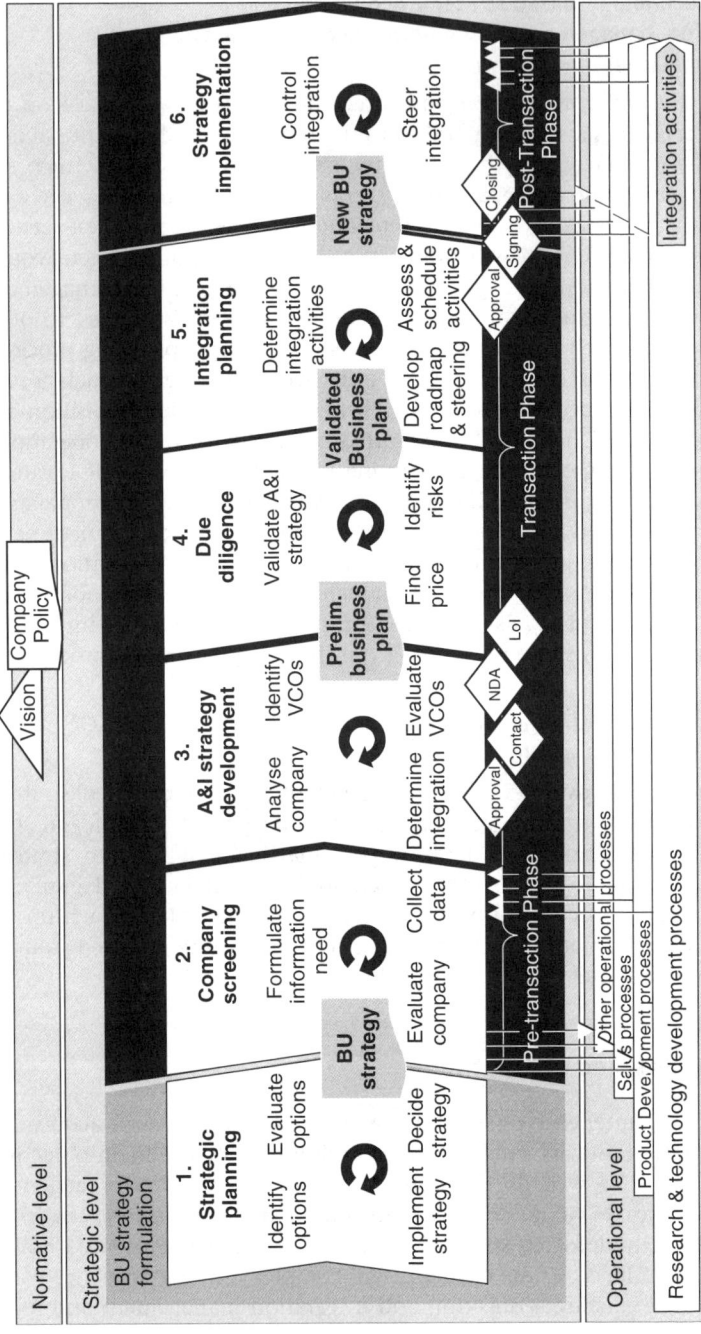

Figure 6.3 A&I process on the normative, strategic and operational levels

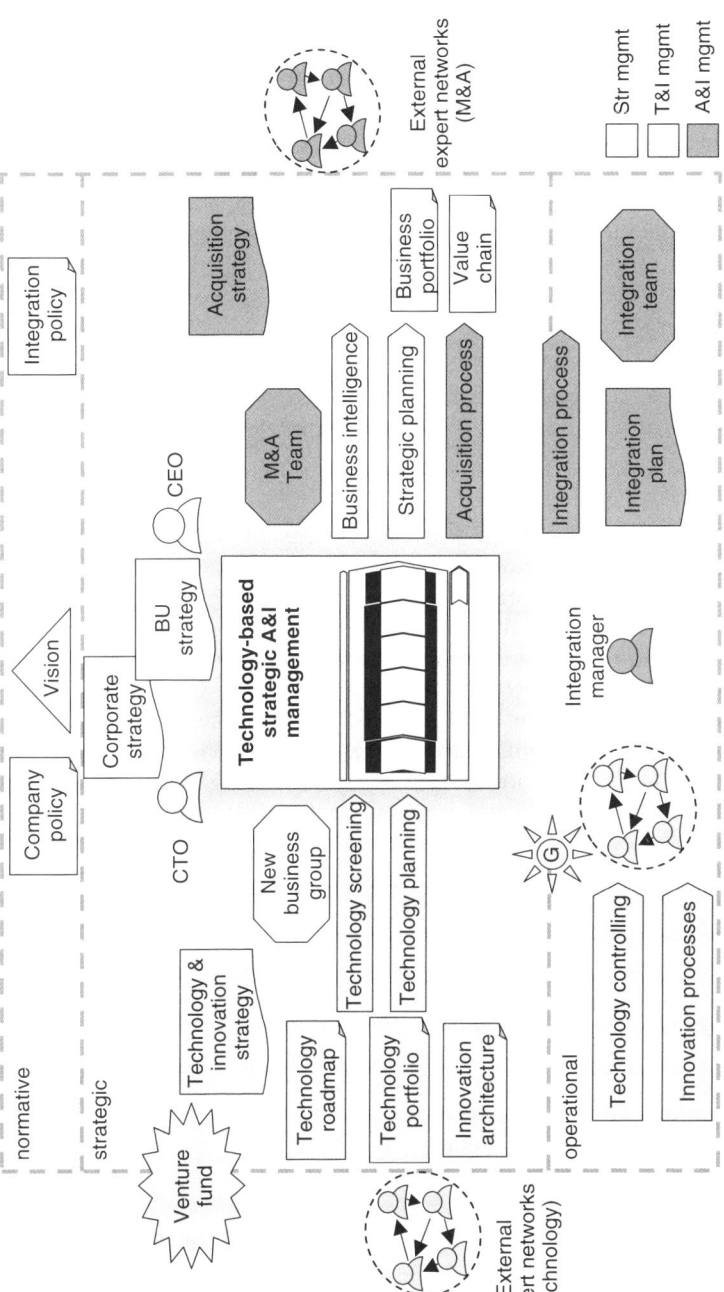

Figure 6.4 Integrating strategic, technology, and innovation, and acquisition and integration management

the occurrence and realization of technological synergies in the course of the integration process. Thus the introduced concept is not a comprehensive guide to overall acquisition management but focuses on the integral elements which support the realization of technology-based value creation. Furthermore, the concept mainly addresses the strategic aspects and thus focuses on strategy formulation and integration planning rather than on their operational implementation. Additionally the concept mainly focuses on the core, the functional and in particular the R&D layer and the integration layer. Tasks of the supportive layer, such as company evaluation, tax, financial and legal issues, IT-related considerations and negotiation strategies, are not addressed as sufficient consideration of there can be found in other books on strategic acquisition and integration management.[2]

6.3.1 Strategic planning

The first phase of the technology-based strategic acquisition and integration management process is the general strategic business planning of a firm. This phase aims to develop a *holistic business strategy* which addresses the strategic targets and paths of all relevant areas of the company. Strategic planning, as an integral part of the acquisition and integration process, further aims to identify strategic growth options for the company which can be pursued via an acquisition. This phase takes place on the management layer of the acquisition only, as the acquisition team has not been established yet. Thus the responsibility for the process lies in the hands of the strategic business planning team and the corporate M&A team led by the business unit head.

To ensure that the identified acquisition options comprise the potential for long-term success in innovation-driven industries and thus enable occurrence of technology-based value creation, strategic planning has to address the following relevant aspects:

- The *acquisition option* has to be included in the strategic planning as a *viable strategic path* to achieve short- and long-term business strategic goals;
- *Technology and innovation strategic planning* has to be an integral part of strategic planning;
- Strategic planning has to consider *internal and external developments* to decide upon the strategic objectives and paths;
- The company's own *ability to pursue the strategic path* of an acquisition has to be considered.

The *generic strategic planning process* consists of four phases: (1) identification of strategic options; (2) evaluation of strategic options; (3) decision upon and formulation of the strategy; (4) implementation of the decided strategy (see Figure 6.5).

Generally, it is recommended that a company be engaged in various different strategic planning cycles with different time horizons.

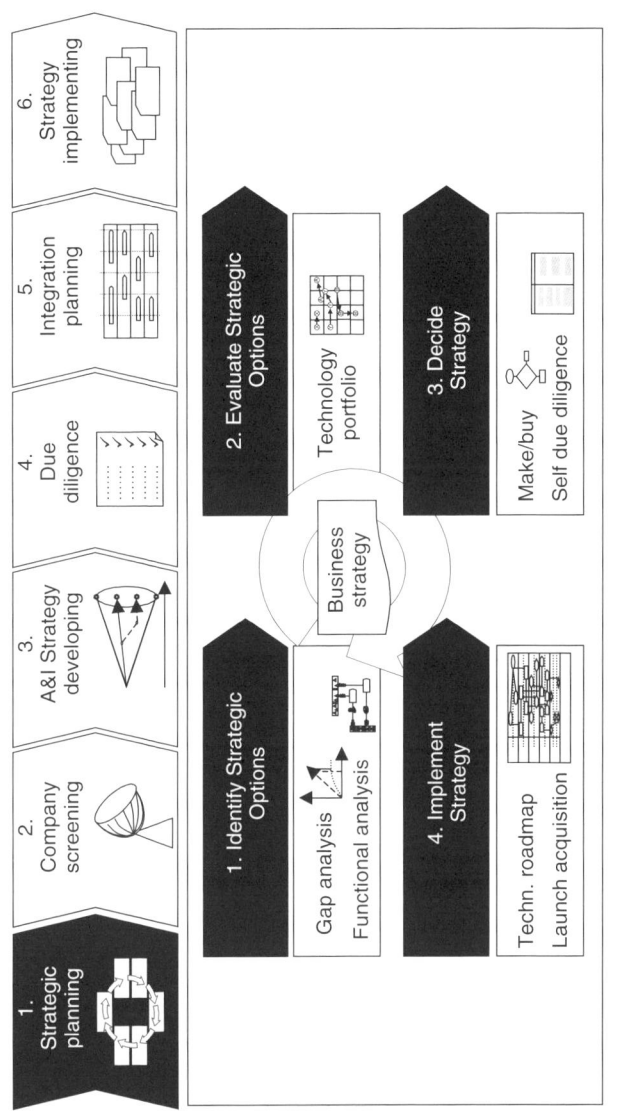

Figure 6.5 Integrated strategic planning

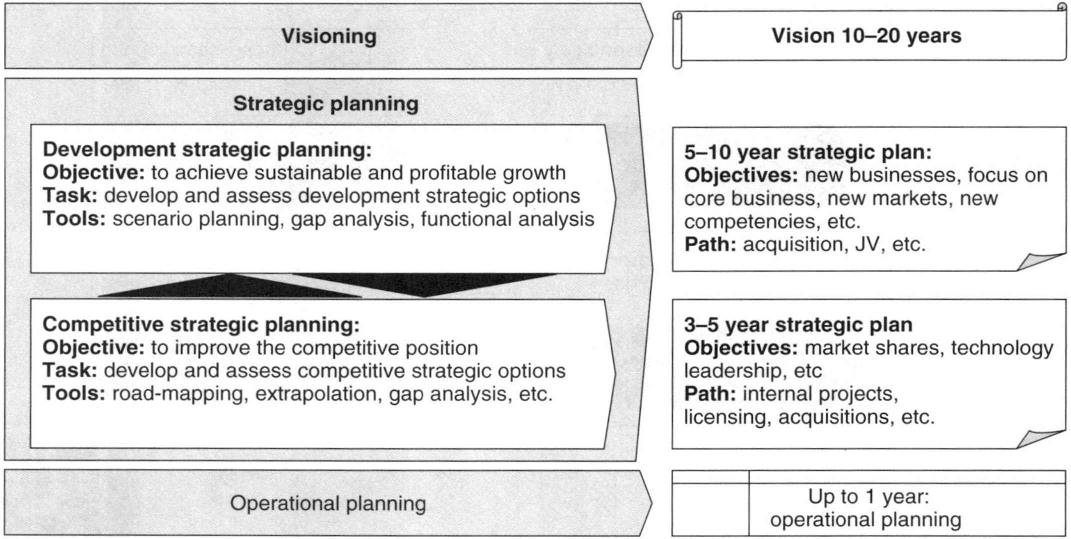

Visioning

Vision 10–20 years

Strategic planning

Development strategic planning:
Objective: to achieve sustainable and profitable growth
Task: develop and assess development strategic options
Tools: scenario planning, gap analysis, functional analysis

5–10 year strategic plan:
Objectives: new businesses, focus on core business, new markets, new competencies, etc.
Path: acquisition, JV, etc.

Competitive strategic planning:
Objective: to improve the competitive position
Task: develop and assess competitive strategic options
Tools: road-mapping, extrapolation, gap analysis, etc.

3–5 year strategic plan
Objectives: market shares, technology leadership, etc
Path: internal projects, licensing, acquisitions, etc.

Operational planning

Up to 1 year: operational planning

Figure 6.6 Planning cycles to allow strategic acquisition

Figure 6.6 provides an overview of the planning cycles on the normative, strategic and operational levels with different time horizons. An acquisition should be considered within the *development and competitive strategic planning* processes. Hence acquisitions which impact on the company for several years can also be strategically consistent. Companies which lack mid- and long-term planning have to address acquisitions more opportunistically, which might negatively affect the long-term performance of the company. To ensure the integration of the acquisition option and the consideration of technological aspects into strategic planning, it is useful to integrate the concepts of *strategic acquisition and integration management* and *technology and innovation management* into the strategic planning process.

Thus the identification of strategic options can be supported by *gap analysis* or *functional analysis*, both methods which can be used to identify a lack of competencies. The strategic options should extend the existing scale and scope of the company (see Figure 6.7), but not address highly unrelated fields in order to avoid unrelated acquisitions. Furthermore, the evaluation of strategic options needs to be enhanced by assessing their technological relevance and timely feasibility, for example by using the *technology portfolio*. Once strategic options have been chosen and gaps have been identified, the question arises of how the strategic objectives shall be pursued. This decision can be supported by *Make-or-Buy* analysis which provides indications of

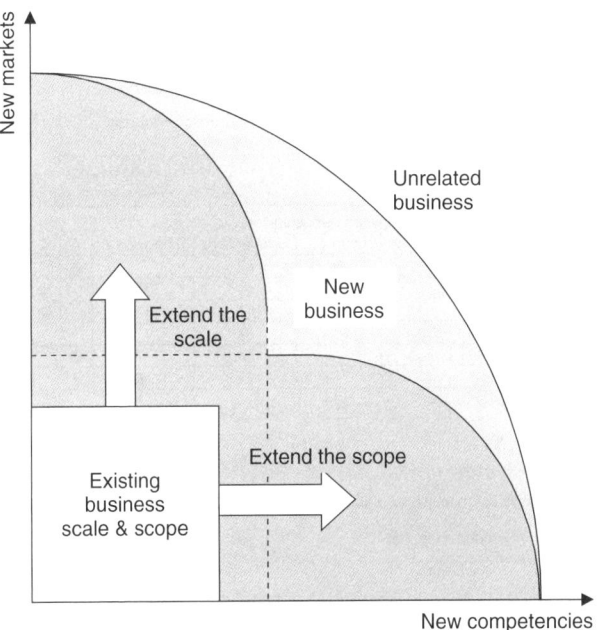

Figure 6.7 Strategic acquisition options which extend scale and scope of the company

whether a strategic objective should be pursued by the company alone or whether the company should cooperate or eventually acquire another company to achieve this objective. Thus the acquisition option becomes one alternative to several strategic paths. A best practice example of how to integrate competence-related aspects and the acquisition option in strategic planning can be seen in Best Practice 3.

Before the decision about initiating an acquisition project is made, the strategic planner should investigate whether the company is able to pursue

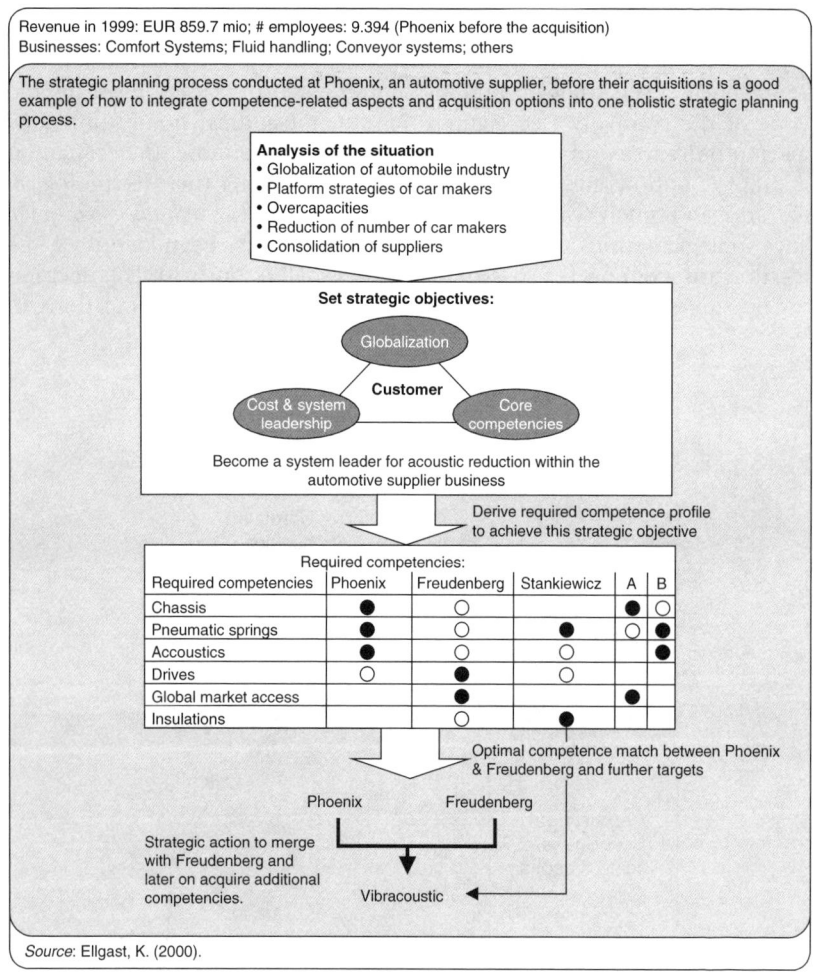

Best Practice 3   Strategic planning at Phoenix

Acquisition Capability	Availability of Resources
Acquisition experience & learning 　　Number of Acquisitions/Year 　　Controlling & learning 　　　mechanisms Degree of formalization 　　Corporate M&A 　　Corporate strategic planning 　　Business development teams Willingness to change 　　Corporate flexibility 　　Open towards change Established technology & Innovation management 　　Clear technology & innovation 　　　Vision & strategy 　　Established technology & 　　　Innovation processes & 　　　structures 　　High innovativeness	Financial resources 　　Financing the acquisition 　　Debt management 　　Financing the integration Skills 　　Leadership & technical 　　　competencies for integration 　　　& knowledge transfer 　　Absorptive capacity Time/Management Attention 　　Management attention & 　　involvement for the whole 　　integration

Figure 6.8 Self due diligence

this chosen path. This investigation of the availability of the acquisition capability can be conducted by a *self due diligence* (see Figure 6.8). This self due diligence is a self check whether the company has the organizational ability and sufficient resources to conduct an acquisition. This self check includes the consideration of whether sufficient acquisition experience has been gained, whether a certain degree of formalized acquisition and integration management and also technology and innovation management is set up and whether the company is willing to change and develop into a new joint company. Furthermore, it is roughly checked to determine if the acquirer can provide sufficient resources in terms of financial liquidity, management attention and time as well as leadership and technological skills to undertake an acquisition. The lack of acquisition competence will foster the awareness of potential upcoming risks. After this self due diligence and the overall strategic agreement to pursue the strategic path of an acquisition to achieve business-strategic goals, an acquisition project should be initiated.

6.3.1.1 Structural and behavioural aspects

From a *structural perspective* it can be recommended that the strategic planning team also includes the CTO and one person from the M&A team, if it exists, besides a person from the business development team. This ensures the consistent integration of technology aspects and the acquisition option in the process. Furthermore, it is recommended that the acquisition team be part of the corporate business development team and not, as is often the case, part of corporate finance. Obviously an acquisition is highly determined by financial aspects but in the end it has to be seen as a strategic path to reach a strategic goal and thus its strategic nature must be in the fore. By combining acquisition and integration responsibility with business development, non-strategic acquisitions can be avoided.

Additionally, the strategic planning process is dependent on internal as well as external information. Thus it should integrate the finding from not only the business intelligence process but also the technology intelligence process and the acquisition screening process, which is often the permanent task of M&A teams and should be an inherent task of the management team. This awareness of technological developments and potential acquisition targets within the same strategic process fosters the potential for realizing a successful acquisition and achieving long-term value creation.

From a *behavioural* point of view, it is important to ensure that the planning process remains a design and rational process and does not become overly political or personal. Furthermore, awareness has to be raised that acquisitions are highly risky undertakings which, once initiated, are often difficult to halt. Thus it has to be ensured that acquisitions are not conducted out of *hubris* or overestimation of own capabilities and that the financial, technological and many other risks are carefully watched. However, in order to reach sufficient momentum to initiate joint willingness to change which is required to achieve acquisition success, the leader of strategic planning has to strengthen the need to pursue the proposed strategic path and thus to create a sense of urgency.

6.3.1.2 Conclusion

It can be concluded that the first phase of the acquisition and integration process has to be the strategic planning process. This has to follow an integrative and holistic approach considering technology aspects and acquisition options at the same time. Furthermore, the process must cover different time horizons and integrate the consideration of the company's own acquisition capability supported by multidisciplinary teams. The result is that the developed growth option pursued with an acquisition is *strategically consistent* with strategic plans, *fits to the technology and innovation strategic objectives* and is understood as one *strategic path* supported by the business. Thus acquisitions which are conducted opportunistically, financially driven and unrelated to the business can be avoided. The acquisitions can change from rarely occurring high-risk projects into an organizational competence.

6.3.2 Company screening

After having identified a strategic growth option which is in line with the business unit and the technology and innovation strategy, an appropriate target company has to be found. This company screening occurs on the management layer of the acquisition and can be conducted in two different ways. It can be understood as a discontinuous process by top management initiated by the strategic planning process or as a continuous task of a corporate M&A or business development team. In this latter case screening runs parallel with business strategy-building and actively includes possible acquisition options and thus widens the scope for corporate development. Thus the screening takes place on the management level of the acquisition. Both the discontinuous and continuous screening activities aim to identify attractive and suitable *potential acquisition targets*.

In order to ensure that the identified acquisition target provides the possibility for the acquirer to achieve short- and long-term and especially technology-based value creation, the screening activities have to address the following aspects:

- The target has to fit to the overall business strategy;
- The contextual aspects of target and acquirer have to fit;
- The target has to provide potential for joint technology-based value creation;
- The target has to contribute also under the condition of the external trends.

The screening process consists of three phases: (1) formulation of information need; (2) data collection; and (3) evaluation of the companies (see also Figure 6.9).

6.3.2.1 Formulate information need

The first phase, the *formulation of the information need*, aims to translate the strategic business objectives from the development and competitive strategic plans into requirements expected from the target company. These requirements should be separated into Need-to-have and Nice-to-have criteria to be fulfilled by the potential target. It is important that the set of requirements not be limited to the actual strategic gaps, such as 'access to the Japanese market' or 'a 3-D laser competence', but cover all strategic areas of the company. This ensures that the target fits with the pursued strategy from all perspectives. It implies that, for example, a company which aims to expand its existing business towards a lower segment requires not only the appropriate market channels but also the cost-efficient design competencies from the target. Thus the larger the acquisition's contribution to business strategy, the more the target has to fit with all areas of the acquirer. Therefore criteria should cover the following areas: market segment and region, product range, core technologies and scientific knowledge, value chain activities, corporate contexts, company value, and financial and legal issues. In formulating the different criteria attention has to be paid so that the focus is on not only the current abilities of the

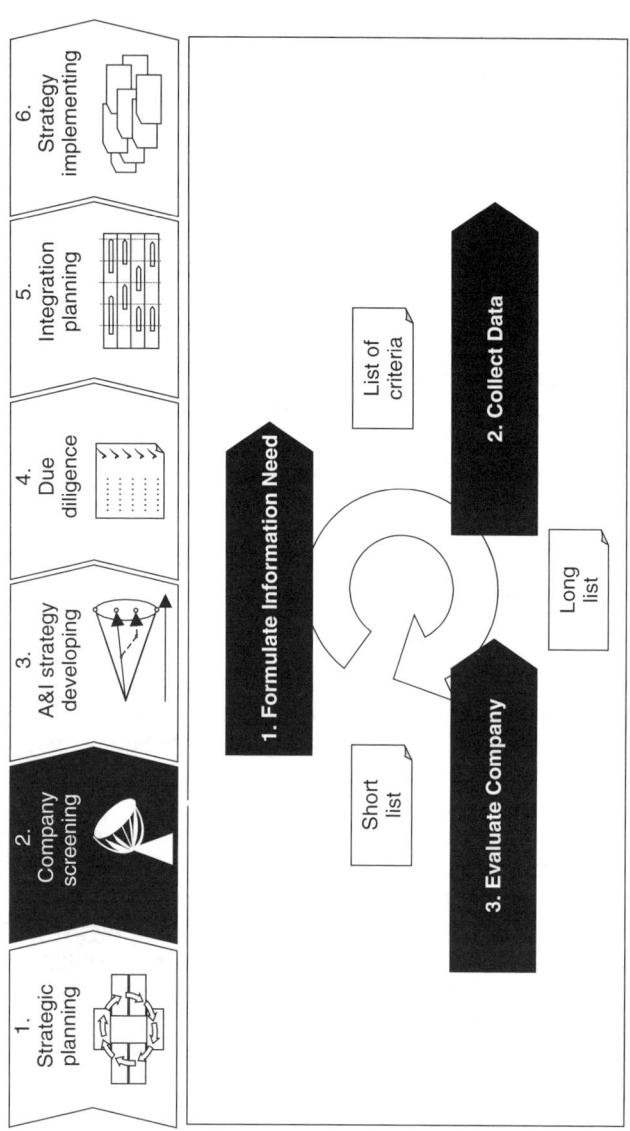

Figure 6.9 Screening process

company but also the ability for the joint companies to create value in the future. This forward-thinking perspective is a crucial element throughout the acquisition process. The acquirer should aim to buy not the most attractive company on the market but the one which best fits and contributes to the joint future strategy. Once the list of need-to-have and nice-to-have criteria is completed, the search for an appropriate target can be initiated.

6.3.2.2 Collect data

Hence the second phase is the *collection of data* about companies. Various sources within and outside the company can be used to retrieve information (see Figure 6.10). In most acquisition management, investment banks and consultants are requested to search for a potential target. Whereas these often have a good overview of the industry dynamics, their focus on technological aspects is quite limited. Thus it seems appropriate to additionally include internal sources of information such as communities of practice, gatekeepers, internal venture funds or the business and technology intelligence teams. If the employees of the company have already been in contact with a potential target, an acquisition can be particularly facilitated as information asymmetry is limited.

As a result of the data collection, a 'Long List' of potential targets can be derived. This needs to be evaluated and checked within the third phase of company screening.

6.3.2.3 Evaluate company

Now the question arises of how the identified potential target should be evaluated in order to arrive at the specific matching target which offers the greatest short- and long-term value creation potential. The answer has already been provided in the previous chapter which outlined the initial conditions required to achieve technology-based value creation. These were, besides the acquisition capability validated within the self due diligence, the (see Figure 6.11):

(1) Value creation potential
(2) Strategic fit
(3) Contextual fit.

At this point of the acquisition process, information asymmetry between acquirer and target is very large. Thus the validation of the required initial conditions cannot be conducted in detail and will be further addressed within subsequent phases of the process. Additionally, the evaluation cannot be conducted within one specific functional area but covers the whole company. Thus the assessments which ensure that the target offers the potential for technology-based value creation are not confined to the R&D area but are more general.

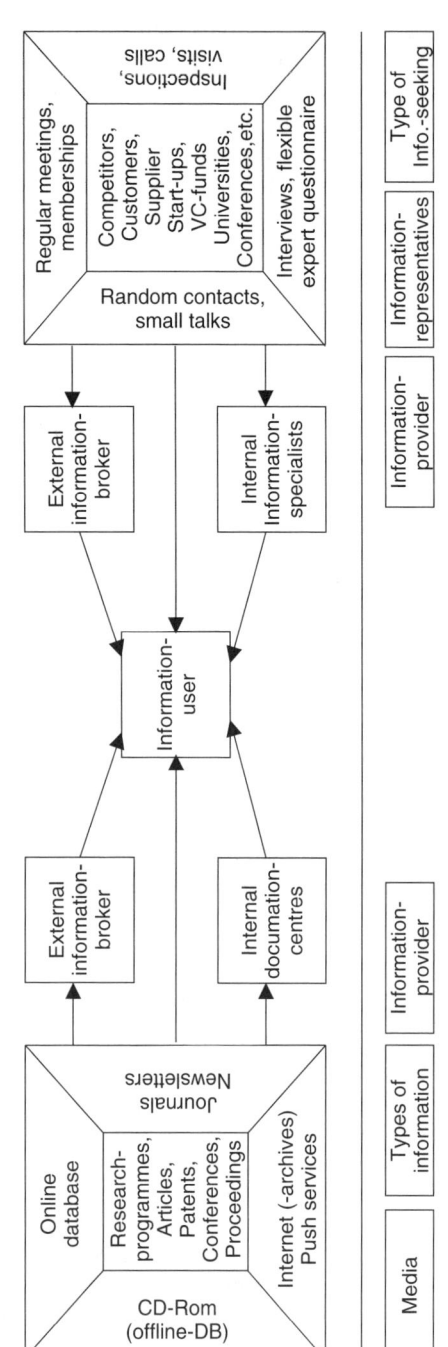

Figure 6.10 Source of information
Source: Lichtenthaler (2000: 38).

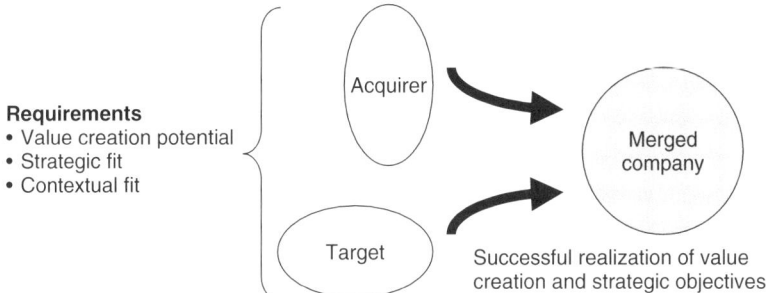

Figure 6.11 Requirements for achieving successful value creation

There are two ways of filtering the potential targets within the long list. In the first step, the list of need-to-have and nice-to-have criteria will be applied to make an initial selection of a potential target. In the second step, value creation potential, the strategic and contextual fit will be assessed individually (see Figure 6.12). Whereas the first filter for addressing the fulfilment of the criteria is quite straightforward to apply, the second individual investigation is more complicated. Currently applied methods to conduct a first rough evaluation, such as Porter's (1987: 46) attractiveness test, cost-of-entry test, and better-off test, or additional checks such as a culture-shock test, verify elements of the above-mentioned criteria but only partly consider technology-related aspects. Thus the following three checks are recommended or should at least be used complementarily.

1. Evaluation of value-creation potential As was outlined right at the beginning of this book, an acquisition always aims for two inherently linked objectives. On the one hand it aims to create value through synergies and on the other hand it should contribute to the strategic path of the business. Whereas the first aspect is considered within this check, the latter is addressed within the subsequent one.

Thus the rough evaluation of value creation potential is concerned with the potential to jointly create value with an acquisition. As the obvious synergy potential is mostly included in the acquisition price and is thus paid for, the real value creation can be derived from only unique, partly hidden and often long-term synergies.[3] Furthermore, long-term value creation in innovation-driven industries is derived from combining and integrating resources and competencies. Thus it is favourable if the future value creation of joint companies is based on the combination of strong resource bases. Additionally, it was observed that there often is a discrepancy between the pursued strategic contribution of an acquisition and the area of actually

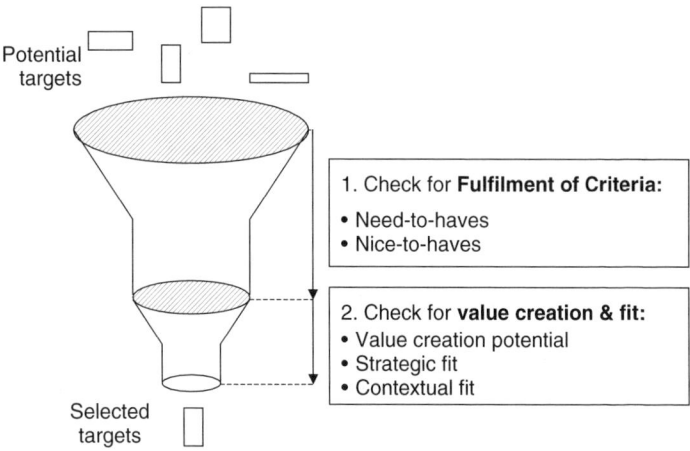

Figure 6.12 Evaluation of potential target companies

achieved value creation. For example, an acquirer aims to internalize technological competencies to increase innovativeness and the main source of value creation also in the long run is derived from head-count reduction. Even though this phenomenon often occurs in the short period after an acquisition, in the long run, the sources of value creation – the synergy areas – should be in line with the strategic objectives of the acquirer. Another aspect relevant to consider regarding synergies is the cost of achieving these synergies. There has to be an appropriate relation between the effort to achieve value creation including associated risks and the returns from these synergies.

Thus the acquirer has to verify that the target company offers potential for synergies which:

- Are inimitable, private and uniquely valuable to the combination of the two companies (similar to the better-off test);
- Are based on the combination of strong resource bases;
- Are in line with the long-term strategic direction;
- Can be achieved with reasonable efforts (similar to the cost-of-entry test).

2. Evaluation of strategic fit The second evaluation of the target is concerned with its overall strategic fit with the acquiring business. Strategic fit is a broad term and needs to be detailed. It implies that the acquirer can achieve its future short- and long-term strategic objectives through the internalization and acquisition of the target company. The different attributes of the strategic objectives determine the dimensions of the strategic fit.

Business strategy	Core elements	Attractiveness	Match to strategy	Fit
Marketing strategy	Core markets	- Size - Growth rate - Competitiveness - Barriers	- Complementary markets - Extended markets - Related markets	Yes/No
Technology & innovation strategy	Core technologies & knowledge	- Maturity - Functional scope - Scalability - Robustness	- Potential to be a core competence - Support of core competence - Fit to technology portfolio	Yes/No
Financial strategy	P/L & balance sheet	- Equity quote - P/e ratio - Cash flow	- Impact on financial leverage	Yes/No
Other strategic areas				

Figure 6.13 Strategic fit in different functional areas

A target thus fits with the acquirer if it:

- Contributes to short- and long-term objectives;
- Matches the pursued strategic goal and path;
- Adheres to corporate and business unit strategy;
- Adheres to the different functional strategies such as marketing, HR, financial and also technology and innovation strategy.

This last match to the different strategic areas is achieved if the target excels within the different areas (similar to the attractiveness test) and thus represents an attractive target from a stand-alone perspective and if it matches the acquirer's intention within this area (see Figure 6.13). Thus, for example, the marketing strategic fit can be achieved if the target serves attractive markets and if these are complementary and important to the acquirer's markets. Whereas at large acquisitions, for example 'Play-for-Scale' acquisitions which highly contribute to the pursued strategy, all company areas should strategically fit, small targets which mainly contribute due to their technology have to fit from only a technology- and HR-strategic perspective.

Fit regarding the technology and innovation strategy can be achieved if the target's core technologies are on the one hand *attractive* and on the other hand *match* the existing technologies of the acquirer. The attractiveness of the core technologies, which at this stage cannot be investigated in more detail, is determined by the acquisition type. When an acquisition aims towards a small start-up company for pace-maker technologies with high potential, it differs greatly from the acquisition of a large competitor which

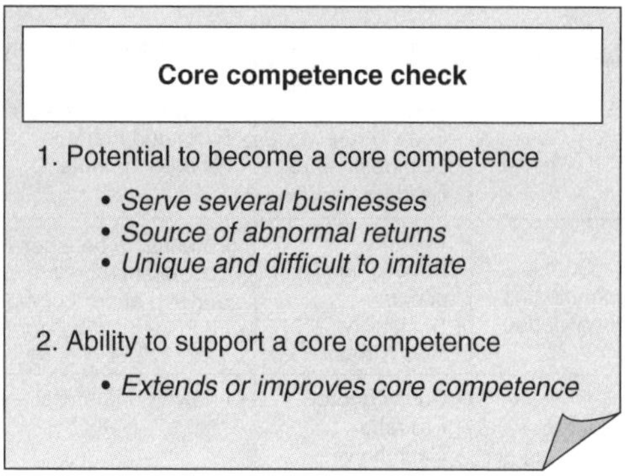

> **Core competence check**
>
> 1. Potential to become a core competence
>   - *Serve several businesses*
>   - *Source of abnormal returns*
>   - *Unique and difficult to imitate*
>
> 2. Ability to support a core competence
>   - *Extends or improves core competence*

Figure 6.14 Core competence check

aims for key or base technologies which can easily be integrated. This matching of the target's core technology can be investigated using two different methods:

- Core competence check
- Technology portfolio.

The *core competence check* investigates whether the target's core technologies have the potential to *become* core technologies of the joined company or if they *support* the core technologies of the acquirer (see Figure 6.14). The rough match to the acquirer's *technology portfolio* also indicates whether the core technologies match with those of the acquirer. The target's attractive technology base and its match with the acquirer's indicate the technology- and innovation-strategic fit of the target.

3. Evaluation of contextual fit Whereas the first two checks investigate whether a target company can be the source of value creation and if it has the potential to contribute to the strategic objectives of the company, this third check focuses on the probability of successfully realizing value creation. Several acquisitions have proven that integration often fails even though the target provides potential for value creation and fits from a strategic perspective. One of the reasons which was found in this research is the lack of contextual fit, which can be avoided by applying the following methodology.

As outlined within the understanding of technology-based value creation within the last chapter, value creation requires the fit of the organizational cultures, structures and dominant business logic of the two enterprises. The amount of fit required depends on the acquisition type and the pursued

> **1. Fit of corporate cultures**
> - Does the target have an attractive, innovation-fostering corporate culture?
> - Does the target's corporate culture match ours?
>
> **2. Fit of organizational structures**
> - Does the target have an attractive, innovation-fostering organizational structure?
> - Does the target's organizational structure match ours?
>
> **3. Fit of dominant business logic**
> - Does the target have an attractive, innovation-fostering dominant business logic?
> - Does the target's dominant business logic match ours?

Figure 6.15 Rough contextual fit evaluation

value creation. Whereas the acquisition of a small innovative company aiming for quick gains does not necessarily require high contextual fit, acquisitions which aim for long-term value creation in turn requiring tighter cooperation and trust, are facilitated by a high but not too extensive contextual fit. Similarly to the other assessments, the contextual fit investigation at this stage can be conducted only on a very rough level, as the company is often not yet personally known. It aims to increase the awareness of potential risks rather than conducting a detailed cultural assessment, which is subject of the actual due diligence phase. In accordance with the strategic fit investigation, the *contextual fit* is determined by the *level of contextual attractiveness* and its *match to the acquirer* (see Figure 6.15). Thus it has to be roughly investigated whether the target masters its organizational structure and culture well and if the dominant business logic including the leadership team follows a professional path. In addition an analysis has to be made of whether the target's context matches that of the acquirer. This investigation includes rumours about the company's business ethic, personal impressions from discussions or role playing and the like. If a contextual fit is already missing from this first investigation, it is recommended that the strategic undertaking is not continued and the focus should move to another potential target.

Active consideration of the value creation potential and the strategic and contextual fit may prevent integration surprises and raise awareness of the upcoming challenges. Best Practice 4 describes Cisco's way of selecting targets.

The screening process ends with the short-list or the identification of a potential acquisition target which is agreed upon by the business unit management and the M&A team. In the next phase this target will be roughly investigated before getting the approval to build a business plan.

Revenue in 2003: USD 18.9 bn; Net Income: 3.6 bn USD; # employees: 40.000
Businesses: Home Networking, IP Telephony, Optical, Network security, Storage networking, Wireless LAN

Cisco is famous for its many and successful acquisitions. At Cisco the acquisition process is part of the business strategy. Cisco buys for several reasons: Shorten time to market, get access to best-of-breed products or technology, expansion into new markets, leverage distribution channel, risk reduction and influx of new expertise (engineering talent).

The screening process is also part of the business process and highly influenced by Cisco's customers. Thus the customers co-determine which company should be acquired.
Furthermore, a 'Culture Cop' ensures the cultural fit of the potential target. Cisco addresses preferably young companies which are or were venture backed and eventually already supported by Cisco (Cisco kids). The screening investigation is supported by methodologies such as role plays which can be used to assess whether the target's strategic decisions would have been made in a similar way by Cisco or a cultural assessment.

Screening criteria:
- Visionary company
- Has a long-term strategy
- Has a finished and tested product
- Has achieved short-term success
- Has cutting edge technology
- Has a deep technical talent
- Geographic proximity
- Preferably privately held
- Is flexible in its way
- Has an equivalent size (mostly 50–100 people)
- Has a similar dominant business logic
- Chemistry and culture fit to Cisco
- Has made a mistake and learned from it

Sources: Anthony, T. K., & Jouret, G. (2000) *Acquisition & integration management at Cisco – Company presentations*; Goldblatt, H. (1999) Cisco's Secrets. *Fortune*: 1–5;
Paulson, E. (2001). *Inside Cisco*. Canada: John Wiley & Sons, Inc;
Holson, L. M. (1998). The Cisco WhizKid: Young Deal Maker Is the force behind a company's growth. *The New York Times on the Web*

Best Practice 4   Company screening at Cisco

6.3.2.4 Structural and behavioural aspects

From a structural perspective company screening is conducted by either the designated M&A team in cooperation with the business unit or top management of the business unit itself. Furthermore, it is supported by internal and external sources of information. If the acquisition need is not explicitly formulated, the screening is often an informal process with little coordination.

The people included in the screening process have to be aware of the long-term risks of strategic acquisitions and thus need to conduct their task very diligently. Furthermore, it is useful if M&A employees participate in the acquisition integration – the most challenging phase of the acquisition. In this way they can learn from the acquisition integration and use their experience to judge at the beginning whether the identified target can be integrated. Additionally the M&A team has to stay in close contact with the business units in order to search for related targets and to ensure strategic fit. Soon after having identified a potential target, the acquisition process is accelerated and becomes more difficult to discontinue.

6.3.2.5 Conclusion

It can be concluded that the screening phase, which aims to identify attractive and matching acquisition targets, consists of three different steps. In the first the need-to-have and nice-to-have criteria which need to be fulfilled by the target are identified. Afterwards the environment is screened for potential targets whereat several internal as well as external sources can be consulted. In a third step the identified targets are roughly evaluated, wherein the focus lies on the future potential joint development of the two companies and on a holistic assessment of the company, ensuring the potential target fits from an overall and thus also from the technological perspective. The assessment is conducted with two different filters. The first checks whether the target fulfils the screening criteria and the second focuses on the individual assessment of value creation potential, strategic fit and contextual fit.

The value creation potential ensures that the joint companies have the potential to achieve short- and long-term value creation. The strategic fit analysis investigates whether the different aspects of the potential target are attractive and match the acquirer's strategy. This investigation takes place in all company areas and thus also addresses the target's match to the acquirer's technology and innovation strategy. Last but not least it is roughly evaluated whether the target's and acquirer's corporate contexts are conducive to value creation and whether they match each other. This also raises the awareness of cultural and organizational differences which can become highly critical issues to master during the integration phase.

This holistic screening ensures that the selected target bears the potential for future joint value creation, fits to the acquirer's strategy and is able to be integrated.

6.3.3 Acquisition and integration strategy development

Once the acquisition target is identified and pre-selected, the subsequent phase, the development of the acquisition and integration (A&I) strategy, is initiated. This phase aims to *develop a business strategy based on the acquisition of the target company*. This business strategy comprises the acquirer's and target's joint strategic goals summarized within the acquisition strategy and paths addressed within the integration strategy. The acquisition strategy contains the objectives to realize joint value creation which support the acquirer in achieving its overall initial business targets, for example the internalization of new competencies. The integration strategy details the company configuration which is required to achieve these strategic objectives. This company configuration is determined by the integration approach and defines the merged company's processes, structures and company behaviour. This joint business strategy, which is similar to a scenario, has to be based on financial figures and forecasts, which are the key indicators in determining the appropriate acquisition price.

Generally the development of the acquisition and integration strategy occurs in *two iterating cycles*, gradually reducing information asymmetry. Within the first cycle of developing an acquisition and integration strategy the acquirer does not have contact with the potential target. Only a small team investigates the feasibility of acquiring the target based on a preliminary business case and the focus is on rough estimations of value creation potential, acquisition price, obvious risks and initial integration aspects. This initial rough strategy development, which is entirely based on assumptions, is used to gain approval for furthering the acquisition process at the board or management team level.

Once the approval to proceed is given, the process gains momentum and becomes more concrete. A core acquisition team is founded, including a transition manager, and supported by a steering committee and functional and supportive subteams or experts (see Chapter 6.3.3.5). The potential target will be contacted and the acquisition intention will be addressed in a very careful and sensitive way.[4] Furthermore, a non-disclosure agreement will be signed to secure privacy. If the potential target is for sale, the subsequent phase of first meetings and company presentations is often substituted with an information memorandum. This second round in the development of an acquisition and integration strategy aims to detail the preliminary business case and underlying financial forecast, validate and revise the initial assumptions while identifying and assessing the value-creation opportunities and determine the integration approach. Furthermore, the direct contact with company representatives helps to get an improved understanding especially of the soft facts influencing the target. This A&I strategy development phase is concluded with a quite detailed future joint business strategy based on assumptions of the potential target and its environment.

Instead of explaining the development of the acquisition and integration strategy twice on two different levels of detail, the process is explained only once and the appropriate application within the two rounds is left to the reader. Generally the acquisition and integration strategy development process consists of four phases: (1) analysis of the company; (2) identification of value-creation opportunities; (3) evaluation of value-creation opportunities; (4) determination of the integration approach (see Figure 6.16). The phases run partly in parallel and overlap with the other acquisition phases. They iterate around the acquisition and integration strategy while detailing assumptions and reducing information asymmetry between target and acquirer.

The development of the acquisition and integration strategy takes place on all layers of the acquisition process. The management layer represented by the core M&A team coordinates the different layers and is in charge of aggregating the different streams of information towards one integrated and holistic acquisition and integration strategy. The functional layers are in charge of developing the functional substrategies. Thus each functional team analyses the company from its own perspective, identifies and assesses value-creation opportunities and contributes to the decision upon the integration approach. The supportive teams aggregate the financial estimations and develop the negotiation and the tax and legal strategies as well as the deal structure. The integration responsible is in charge of developing an integration approach and configuration with the support of the functional and supportive layers. However, as the tasks of the supportive subteams are extensively addressed in current literature[5] and have no impact on technology-based value creation, this layer is not outlined within this book. The focus rather is on the R&D functional layer, the core M&A team and the integration team.

In order to ensure the appropriate consideration of technology aspects within the subteam developing the technology- and innovation-related acquisition and integration strategy, the introduced understanding of *aspects relevant for technology-based value creation* implies the following:

- The development of A&I strategy has to consider the implications from the different acquisition types for the potential realization of technology-based value creation;
- A&I strategy has to fit with the overall strategy;
- The development of A&I strategy has to include the prospective and integrative consideration of technology-based value-creation opportunities;
- The development of A&I strategy has to determine the appropriate integration approach for achieving technology-based value creation.

6.3.3.1 Analyse the company

In the first step of developing the technology- and innovation-related acquisition and integration strategy, the target has to be analysed from a

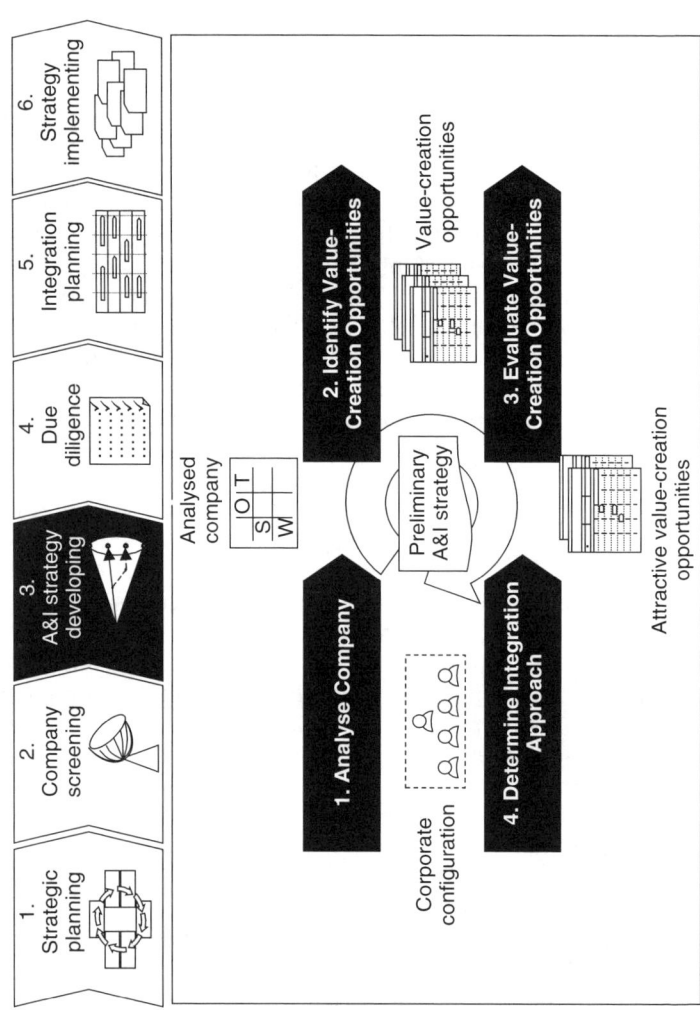

Figure 6.16 Development of the acquisition and integration strategy

	Target's resources	Target's context
Objectives of analysis	– Get an overview – Identify strengths & weaknesses – Identify trends & developments	– Get an overview – Identify strengths & weaknesses – Identify trends & developments
Elements of investigation	- Markets, customer needs - Products, modules - Product & process technologies - Knowledge - Business fields, functions & technology platforms - Linkages & interdependencies	- R&D Organization & processes - Innovation culture - Technology & innovation strategy - Management's ability
Applicable methods	- Innovation architecture - Technology roadmap - Customer needs analysis - Value chain analysis	- Organizational charts - Process flow charts - Contextual web

Figure 6.17 Objectives, elements and methods of a technology and innovation analysis

technology- and innovation-related perspective according to its strengths and weaknesses and the upcoming trends which it will face. It is important to mention that the subsequent analysis partly overlaps the analysis of the target in various other fields and can thus be conducted in cooperation with other teams analysing the potential target.[6] This analysis addresses two main areas: (1) the target's resource perspective comprising the markets, products, modules and product and process technologies, and the underlying knowledge; and (2) the technology- and innovation-related corporate context such as the R&D organization and processes, culture and dominant logic and the R&D management team. This analysis aims to collect data required within the subsequent steps of the acquisition and integration strategy development and the technology due diligence and is supported by various tools (see Figure 6.17). The technology-related analysis of the strengths and weaknesses as well as trends and developments is conducted with methodologies of technology and innovation management applied by the technology and innovation team or the technology-responsible person at the acquisition. This is often the CTO of the acquirer or the head of engineering.

1. Analysis of the target's resources The first technology-related analysis of the *target's resources* has to address the following aspects:

- What are the target's resources and product and process technologies?
- What are the target's core competencies?

- What are the target's strategic technology platforms, functions and underlying knowledge?
- What are the target's business fields, markets and customer needs?
- How are the resources integrated in the current products and linked to each other?
- How are the technologies deployed within the value chain?
- How are resources planned to evolve and change over time, including an investigation of what the target has in its product and technology pipeline?
- Where are the main strengths and weaknesses?
- What are the main trends influencing the resource base?

All these aspects can be considered by developing an innovation architecture, a roadmap, a value chain analysis and a customer needs analysis for the target company. The innovation architecture,[7] for example, shows which business fields, functions and strategic technology platforms the target has and which resources are so far mastered, which are planned or not yet well mastered (see Figure 6.18). The level of mastery should be roughly estimated by the expertise of the target within the areas, the amount of associated specialists or number of patents within the field. Furthermore, it shows the current application of the different resources to achieve technology-based value creation and can be extended by upcoming customer needs. The innovation architecture can also be analysed for core competencies. An associated roadmap of the target company shows the ongoing and planned development projects.

The innovation architecture mainly focuses on the interrelation of resources to achieve new product innovations. The perspective of the technology usage within the value chain as production and engineering process technologies is addressed in only a very limited way. Nevertheless, applied and newly developed process technologies within the value chain are often an attractive source of technology-based value-creation opportunities, especially resource-deployment opportunities, or associated with specific risks, such as the difficulties of transferring CAD technologies. Thus the analysis should also be concerned with the strengths and weaknesses, trends and developments of the process technologies within the different value activities.[8]

2. *Analysis of the target's context* The second analysis of the target company is concerned with its *technology- and innovation-related corporate context*. Thus the investigation has to address the target's *R&D structure and process*, the *innovation culture* and the technology- and innovation-related dominant business logic, including the ability of the *R&D management team*. This analysis should roughly indicate whether the target's corporate context is conducive to technology-based value creation. It can be measured by applying the methodology introduced in Figure 6.19. The attributes within the analysis indicate whether the corporate context is conducive to innovativeness (high score) or not (low score).

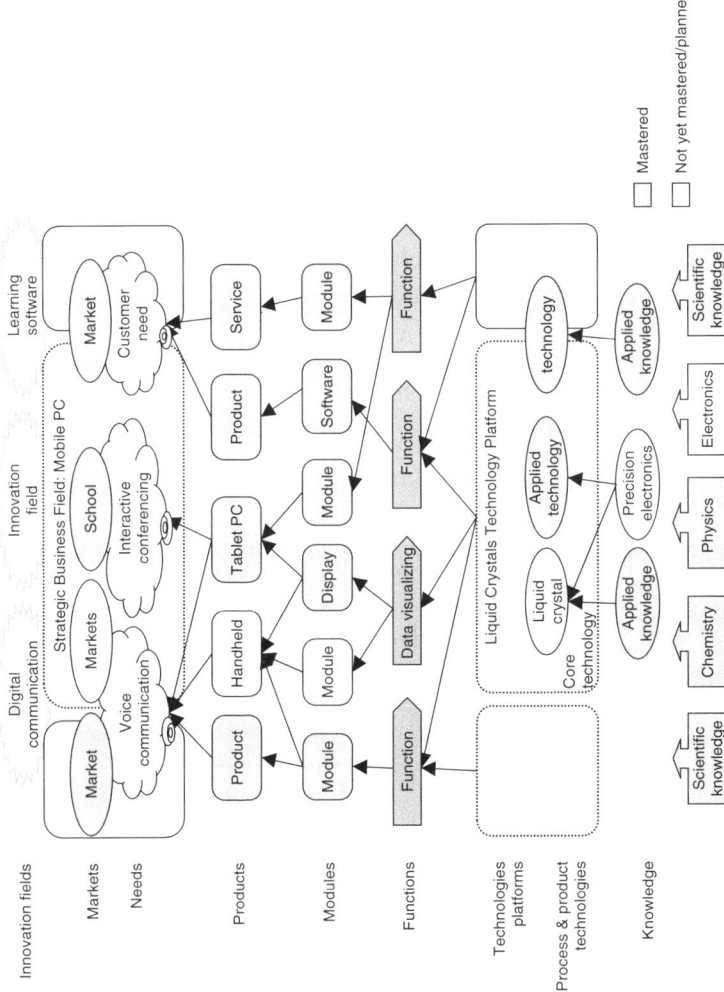

Figure 6.18 Analysis of the target's resources
Source: Sauber (2003).

Context	Characteristic		1 --------- 5		Acquirer	Target
Cultural context	Openness	Closed	vs.	Open	3	4
	Leadership	Technocracy	vs.	Entrepreneurial	4	2
	Time	Competitive focus	vs.	Development focus	2	3
	Reaction to change	Stability	vs.	Flexibility	3	1
	Diversity	Uniform	vs.	Diverse	4	4
Dominant logic context	Technology strategy	Follower	vs.	Leader	3	3
	Value driver	Efficiency / Cost	vs.	Value propositions	4	5
	Technology application	Scale	vs.	Scope	3	3
	Technology marketing strategy	Make & Keep	vs.	Buy & Sell	2	4
	Innovation rate	Small	vs.	Large	5	3
Organizational context	Configuration	Centralized	vs.	Decentralized	3	2
	Orientation	Task	vs.	Individual	5	5
	Boundaries	Low connectedness	vs.	High connectedness	2	4
	Level of autonomy	Heteronomy	vs.	Self-determination	4	3
	Level of formalization	Bureaucratic decision making	vs.	Formalized flexible decision making	5	2

Figure 6.19 Attributes of the target's corporate context

Sources: Schein (1992), Bleicher (1999) and Pumpin (1985).

The ranking should be related to the acquirer's corporate context. The indicators are often soft rather than hard facts and based on personal impressions and feelings and thus need to be analysed in the course of the investigation. This analysis of the target's weaknesses and strengths and especially of its future developments from a technological perspective needs to be integrated into the overall calculation of the acquisition price conducted by the financial team, usually headed by the CFO. This team can either evaluate the stand-alone company value only roughly via using multiples of companies within a similar business area or with similar technologies, or more deeply investigate the acquisition price based on the future undertakings of the target by using NPV methodology.

The phase of analysing the target within two rounds conducted via interviews, presentations and the like results in a good overview of strengths, weaknesses and developments of the technology and innovation aspects of the target's resources and corporate context. If the analysis results in findings of specific weaknesses, then these have to be marked as potential threats. One best practice is a flag methodology, within which the high risks which have the character of deal breakers are marked with a red flag and stay pending throughout the acquisition. Medium risks are marked with yellow flags and resolved issues turn green. This methodology, which can be applied during the whole acquisition process, is easily applicable and provides an overview of the risks.

The findings of the analysis will be used within the subsequent phases to identify and evaluate value-creation opportunities and to conduct the technology due diligence.

6.3.3.2 Identify value-creation opportunities

After having identified the strengths and weaknesses of the target and the trends impacting on it, the value creation potential of the joint companies has to be identified. This comprises the identification of value-creation opportunities within the whole company, such as in the sales and marketing areas as well as in the technology and innovation area. Within this technology and innovation layer, the identification of value-creation opportunities is identical to the development and revision of the joint technology and innovation strategy and plan. Thus the future product and technology pipeline for the joint company is being developed. Therefore the term value creation opportunity is used deliberately as opposed to the term synergy as the identification of new value creation options covers synergistic value creation as well as value creation from the stand-alone companies.

Within this technology and innovation area two different value-creation opportunities can be identified: innovation and resource deployment possibilities. Furthermore, short-term can be distinguished from long-term opportunities.[9] Whereas the first mainly focus on cross-selling opportunities or the leveraging of modules or the like, the latter focuses on competence-building measures and long-term innovativeness. This potential to achieve long-term value-creation opportunities is difficult to assess as mostly the acquirer has no detailed plan of long-term product developments, etc. Nevertheless, rough considerations on joint future innovativeness and the deployment of resources help to assess the value of the acquisition and raise awareness of upcoming risks and challenges.

Before starting to identify technology-based value-creation opportunities, a new awareness of the companies' resources has to be developed (see Figure 6.20). The focus of identifying value-creation opportunities should not be confined to finding some possibilities for improving a product but should take advantage of a newly merged, enhanced and extended technology base available to target and acquirer. Thus the newly formed company's innovativeness can be based on this much broader technology pool. Furthermore, the enhanced technology base offers the potential for improving efficient deployment of existing capabilities and for building new ones. This awareness can be fostered by introducing the same terminology and concepts to master the newly merged resource bases. Thus the different technologies of both companies can become part of the same strategic technology platform and fulfil the same functionality. The CTO should take a leading role in defining a joint technology management terminology. A similar understanding is pursued by Cisco; see Best Practice 5.

Generally several different ways of identifying value-creation opportunities can be distinguished. In the following text two different but interrelated approaches which help to identify joint technology-based value-creation opportunities will be shortly introduced. The first is concerned with the identification of innovation opportunities and the second is concerned with the identification of resource-deployment opportunities (see Figure 6.21).

1. Identification of innovation opportunities As described, innovation is concerned with the integration or reconfiguration of resources into new products, services or product platforms which, once successfully introduced to the market, become innovations. As mentioned, innovation is driven by two different factors, market pull, seen as the customer's demand for improved or new functionality, and technology push, which is the company's urge to integrate new technologies in the products. According to these drivers there are two different possibilities of identifying innovation opportunities: (1) market pull analysis; and (2) technology push analysis.

The identification of *market pull* innovation opportunities is initiated by the investigation of upcoming customer trends within the markets of the target and the acquirer. These trends or customer needs can also be translated into the need for a new or revised product with certain functional requirements (see Figure 6.22). Subsequently, an evaluation has to be made of how the different customer needs can be satisfied by the newly merged company. This analysis can help identify cross-selling opportunities or the need for developing a new product or product platform. Furthermore, the

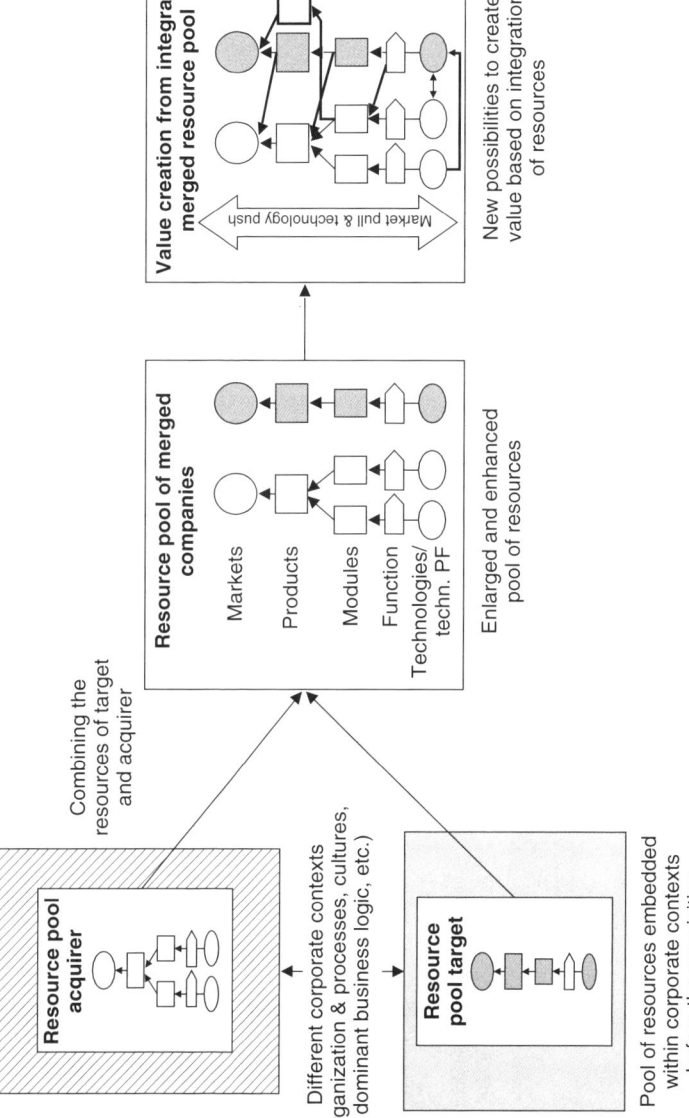

Figure 6.20 Merging and integrating the target's and acquirer's resource bases

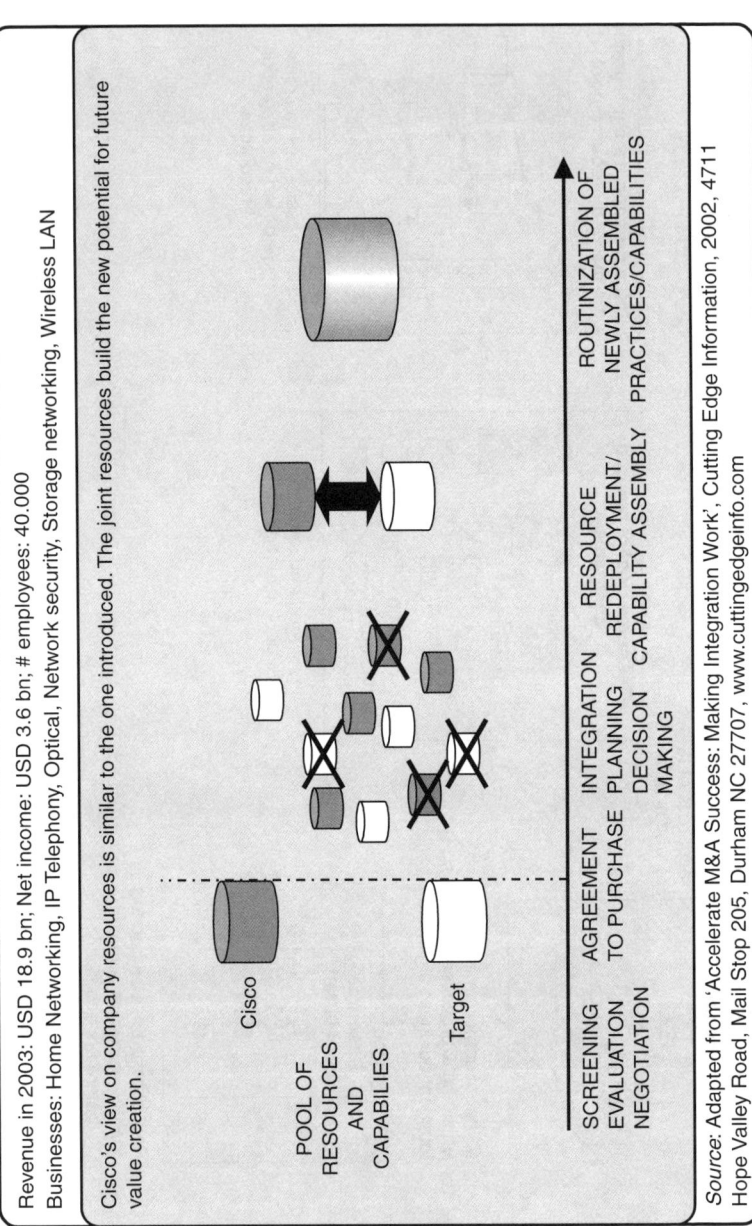

Best Practice 5   Understanding of merging resources at Cisco

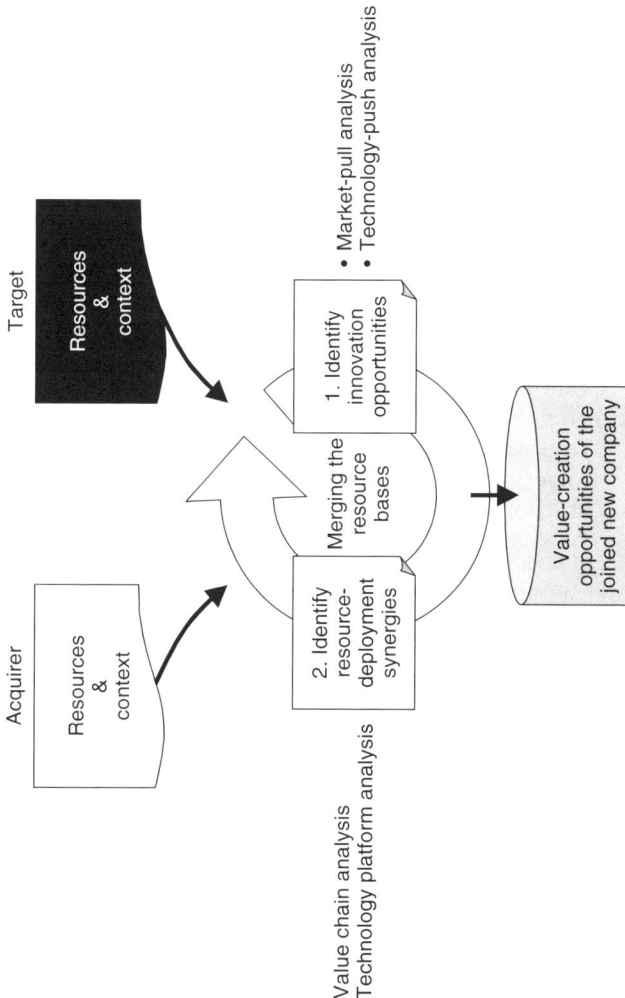

Figure 6.21 Identification of technology-based value-creation opportunities

possibility of leveraging technologies or knowledge can be identified by asking which of the technologies for a specific strategic technology platform of the target and the acquirer or of an external third party best fulfils the functional requirements defined by the customers. The results of the analysis are innovation ideas for the next product revisions and generations.

Technology push analysis to identify innovation value-creation opportunities is complementary to market pull analysis and can in the best case arrive at similar results. Whereas market pull analysis focuses on satisfying upcoming customer needs, technology push analysis focuses on the integration of attractive technologies. Thus it starts with the investigation of the technology bases of target and acquirer and the identification of attractive resources. The technologies are categorized within strategic technology platforms which provide a functional scope with a certain performance. Subsequently, combining these attractive resources to use the functional scope for creating customer value is considered (see Figure 6.22). This bottom-up analysis can arrive at innovative ideas about how to leverage a technology or module to enhance or extended provided functionality or to develop a new product platform based on the reconfiguration of existing resources.

In identifying innovation opportunities, different time horizons should be distinguished. Thus easily achievable short-term value-creation opportunities should be considered as well as more challenging joint platform developments which will be initiated much later after the acquisition.

2. Identification of resource deployment synergies Besides the identification of innovation possibilities, opportunities for redeployment of the newly merged resource bases have to be identified. Resource-deployment opportunities are the efficient transfer, or substitution, of resources which results in an efficient redeployment of capabilities and thus in cost and complexity reduction. Often resource deployment is initiated by related innovation activities. For example, the joint development of a new product might require co-location of the engineering teams or the transfer of R&D process technologies. Nonetheless resource deployment can also occur independently from innovation. Two different methodologies for identifying resource-deployment opportunities will be introduced: (i) value chain analysis; and (ii) technology platform analysis. Whereas the first focuses on the deployment of process technologies, the second supports the distribution of product technologies within the joint company.

The concept of the *value chain*, initially introduced by Porter,[10] is a very useful tool for identifying synergies in all areas of company activity. Thus it is used for the identification of operational synergies and the like. Within the identification of technology-based value-creation opportunities, value chain analysis can be used to search for opportunities to transfer, fuse or substitute technologies or resources between the two companies in order to increase the performance of the activity (see Figure 6.23). This analysis

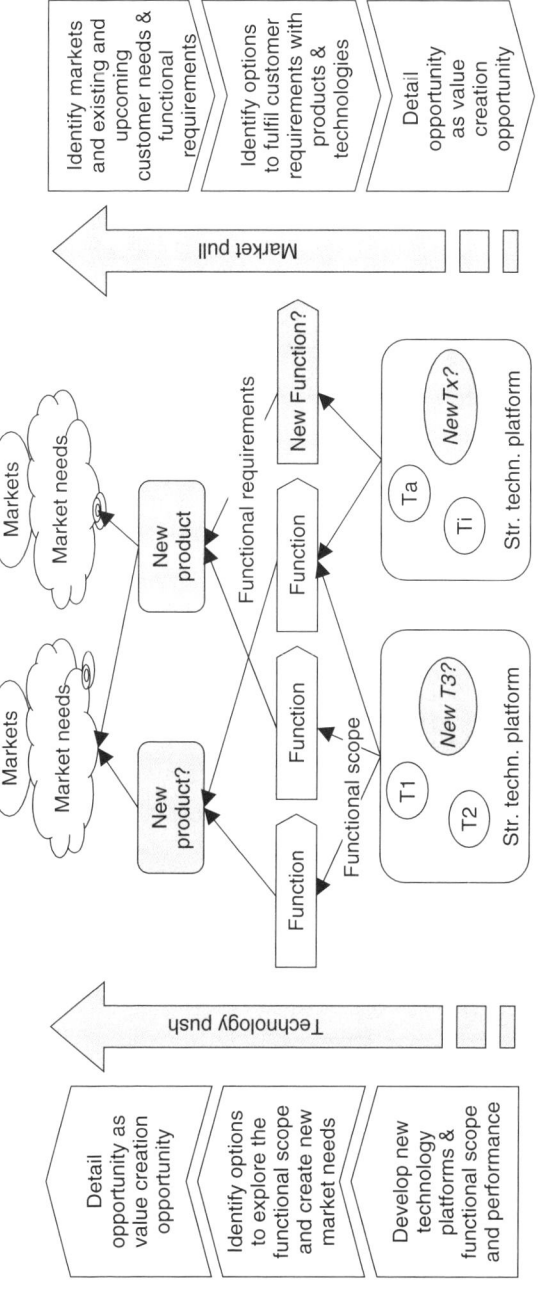

Figure 6.22 Market-pull and technology-push technology-based value-creation opportunities

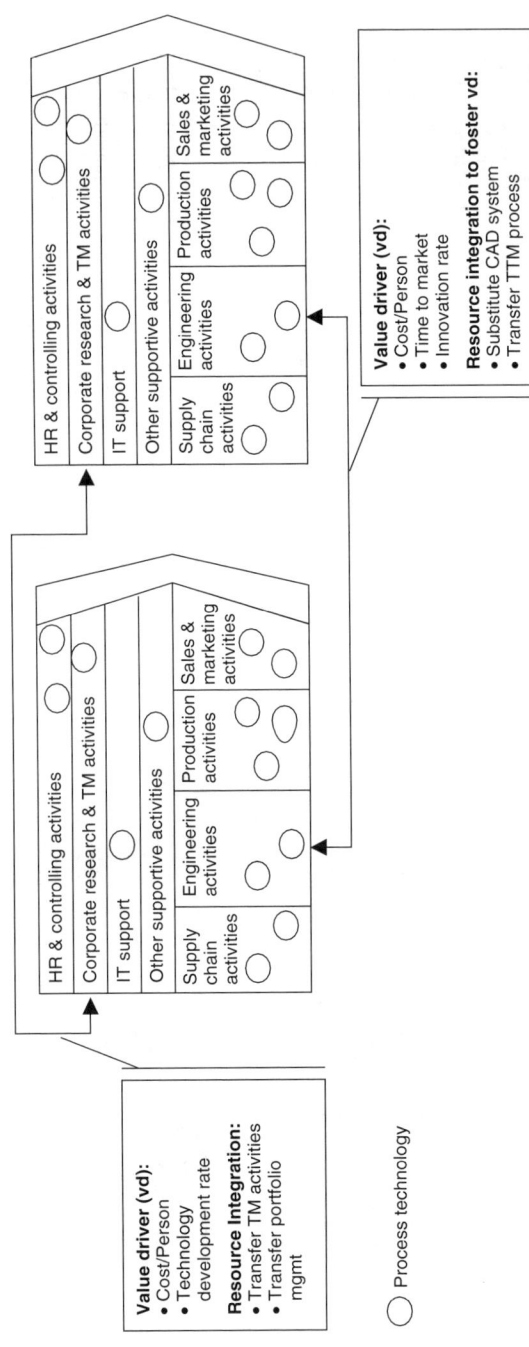

Figure 6.23 Value chain analysis for the identification of resource deployment synergies

focuses on the R&D area but it could also be applied within all other value chain activities.

Thus in a first step the value chains and the process technologies applied within them have to be identified. Additionally, the value drivers for each activity, such as the cost per unit or per person, have to be added. Then opportunities to transfer, substitute or fuse technologies which affect the value drivers have to be identified and subsequently be detailed as value-creation opportunities. The second step in identifying resource-deployment opportunities focuses on not applied process technologies but the efficient deployment of product technologies within the joint *strategic technology platforms* and within the *joint companies*. There should be an attempt to optimize the technologies within each platform according to the functional requirements over time and according to the life-cycle management of the technologies. Furthermore, the technologies should be efficiently deployed within the whole company, aiming to reduce redundancies and to foster the bundling and building of competencies according to needs (see Figure 6.24).

Thus it is proposed first to identify the common strategic technology platforms within the companies and the technologies of acquirer (T(a)) and target (T(t)) (see Figure 6.24) within these. Additionally the functional requirements derived from the identification of innovation opportunities need to be summarized in a product and technology roadmap. Subsequently, each strategic technology platform needs to be optimized according to these functional requirements and to the life-cycle management, for example by using a technology portfolio. Furthermore, the technologies within the platforms should be deployed within the whole company to avoid redundancies and to foster the bundling and building of competencies, for example by building competence centres. The main difficulty then lies in the definition of the value creation opportunity with an associated potential value creation expressed in financial figures. As very often these types of value-creation opportunities have no direct but only an indirect effect on value creation, their contribution either needs to be split over other value creation projects or should be neglected if the figure loses credibility.

Once the value-creation opportunities are identified, they have to be detailed and explained as such. Besides being identified with a name and later on a responsible person, a value-creation opportunity can be described as a sum of interconnected development projects which aim to achieve an innovation. Thus a value-creation opportunity can comprise technology development projects, module and product development projects, marketing projects, the adaptation of the production processes and also a period where it is sold to the market. Additionally, each value-creation opportunity can be assigned a value driver and appropriate objectives such as a pursued sales volume or a rough net present value estimation (see Figure 6.25).

Within the acquisition and integration process only the most important value-creation opportunities will be focused on and further developed

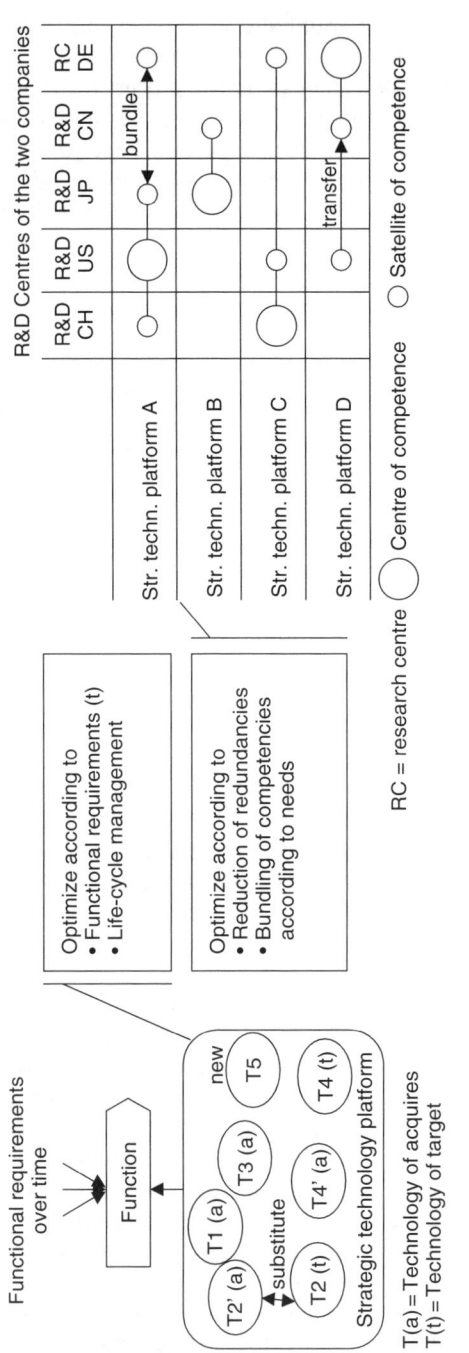

Figure 6.24 Optimization of technology deployment

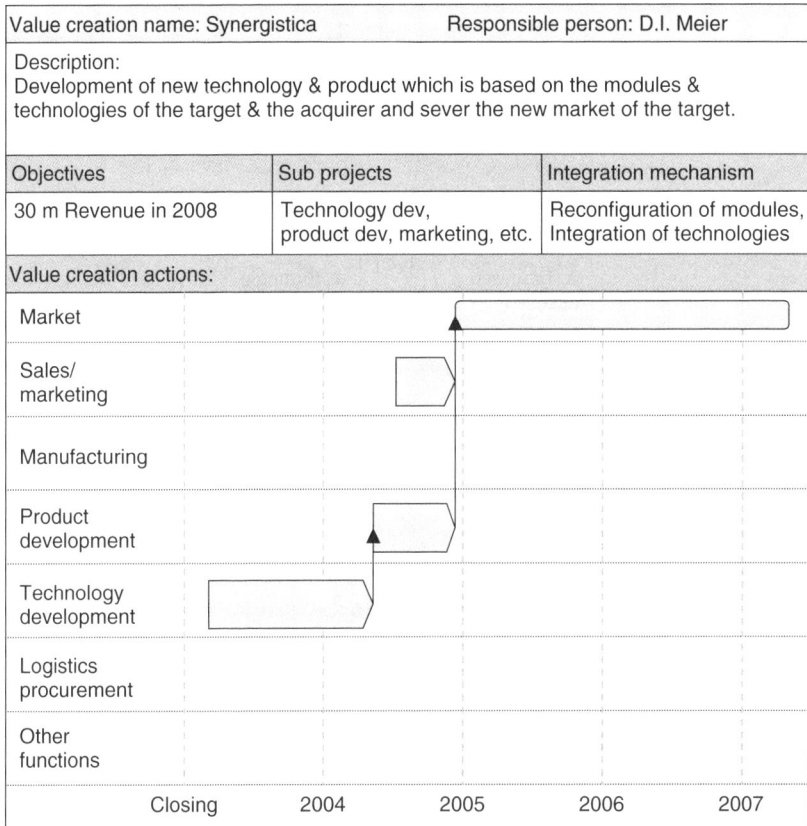

Figure 6.25 Documentation of value-creation opportunity

within the business plan, evaluated to calculate the acquisition price, validated within the technology due diligence and later on realized within the integration phase. Thus the large pool of identified value-creation opportunities needs to be narrowed down to the most important options by evaluating them.

6.3.3.3 Evaluate value-creation opportunities

The evaluation of the value-creation opportunities comprises four steps: (1) evaluation of strategic fit; (2) evaluation of financial profitability; (3) evaluation of the associated risks; and (4) evaluation of the timeliness of the value-creation opportunity (see Figure 6.26). These assessments will limit the amount of pursued value-creation opportunities and ensure their attractiveness within

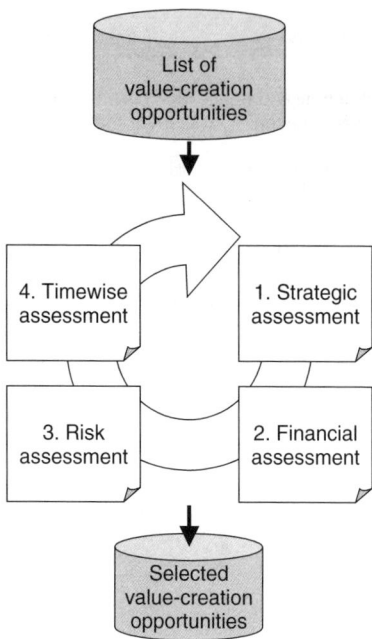

Figure 6.26 Evaluation of value-creation opportunities

the acquisition. They differ from the rough company-focused evaluations within the screening phase as within this phase the focus lies on individual value-creation opportunities which can be understood as projects.

1. Strategic assessment First of all the strategic fit of the value-creation opportunities is evaluated. This investigation is adapted from Porter (1996) who distinguishes between three different orders of strategic fit. The first-order fit describes the fit of the activity with the overall strategy. The second-order fit describes the consistent reinforcement of the activity of other activities. The third-order strategic fit goes beyond the reinforcement of other activities towards the optimization of the overall company's efforts, which thus relates to a fit of the activity with the overall company configuration (see Figure 6.27). Thus the assessment of the strategic fit of a value-creation opportunity comprises the following considerations. Does the technology-based value-creation opportunity:

- Fit with the overall corporate and business unit strategy (foster objectives, etc.)?
- Fit with the technology and innovation strategy (in line with the technology portfolio and core competencies, etc.)?

- Fit with the competitive and development strategic plans (contribute in the short and long run to the strategic objectives, etc.)?
- Reinforce the other value-creation opportunities and further integration activities (technological synergies, etc.)?
- Reinforce the daily business (support sustaining value)?
- Optimize the effort within the company organization and processes (can be managed with the processes, etc.)?
- Optimize the efforts within the company culture (match the company culture)?
- Optimize the efforts within the dominant business logic (match to value drivers)?

2. *Financial assessment*   The second assessment evaluates the value-creation opportunities from a financial perspective. Thus only value-creation opportunities which contribute to the company value and thus have a positive NPV (net present value) are attractive to pursue. Even though the financial figures are often difficult to attribute to technology-related projects or resource-deployment activities, a rough estimation has to be attempted in order to ensure the appropriate consideration of the value-creation opportunities within the price calculation and to support the guidance and controlling of such a value-creation opportunity. Figure 6.28 provides an overview of the calculation method applied in calculating the NPV. The financial assessment of value-creation opportunities within acquisitions has to integrate cash flow

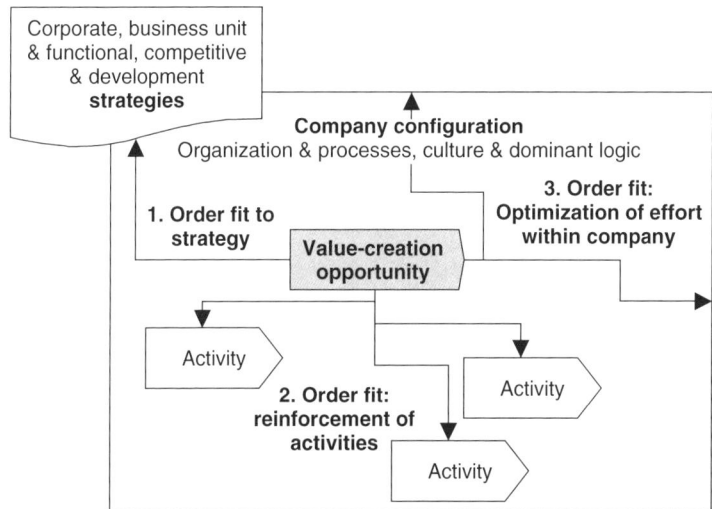

Figure 6.27   Three order fit paths
Source: adapted from Porter (1996: 71ff).

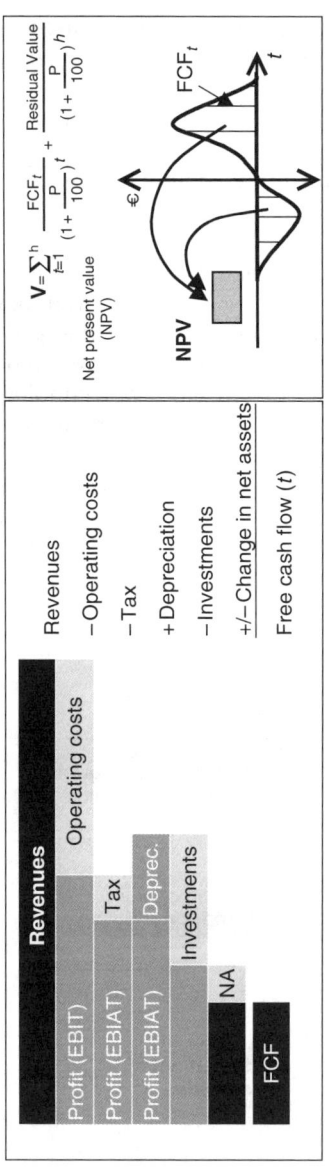

Figure 6.28 Net present value calculation

streams from the associated integration and restructuring effort as well as from the synergy effects. This ensures the appropriate consideration of research and development efforts within the strategic acquisition rationale and calculation.

3. Risk assessment The third evaluation of the value-creation opportunities relates the calculated NPV to the risks associated with the opportunity. Generally internal and external risks can be distinguished. The internal risks concern the technological risk, risks associated with the availability of resources required to realize the project and risks which occur with the specific characteristics of the acquisition. For example, a hostile takeover of a highly unrelated company in another cultural and geographic area increases the risk level of the value-creation opportunities. The external risks associated with a value-creation opportunity come from the market side, such as customer and supplier behaviour, from the competitors' side, such as substitutive or new competitors, and from other areas such as legal, environmental or political risks. As outlined in Figure 6.29, value-creation opportunities with a high net present value and low risk should be pursued, whereas if the risk is also high the acquirer should carefully ponder which risk level it intends to pursue. Similarly, the acquirer should ponder or omit value-creation opportunities with low NPV and low associated risks, whereas value-creation opportunities with low NPV and high risks should be omitted if possible.

4. Timely assessment The last assessment of the value-creation opportunities concerns their timely occurrence. As was outlined, acquisitions of highly innovative companies aiming for internalizing competencies are often under high pressure to innovate. Thus acquisitions are used to quickly access a new market or to increase the innovation rate. Thus the question of whether the identified value-creation opportunities can be realized on time has to be evaluated. This evaluation can be conducted based on the roadmap of the different value-creation opportunity projects (see Figure 6.30). Whether the technology, product or marketing activities can be coordinated to meet the market demand on time is assessed. For example, if the core technology of the target is not ready early enough to serve the market before the competition does, the whole acquisition logic has to be questioned.

Following the value-creation opportunities' strategic, financial, risk and timeliness assessment, only a limited number of technology-based value-creation opportunities remain. These dominate technology-related considerations within the subsequent phases of the acquisition and integration processes. They determine the integration approach, are the focus of the technology due diligence and the main activities to plan and address during the integration phase.

6.3.3.4 Determine integration

Within the last two sections of this chapter the acquisition strategy, including the identification and evaluation of possible value-creation opportunities,

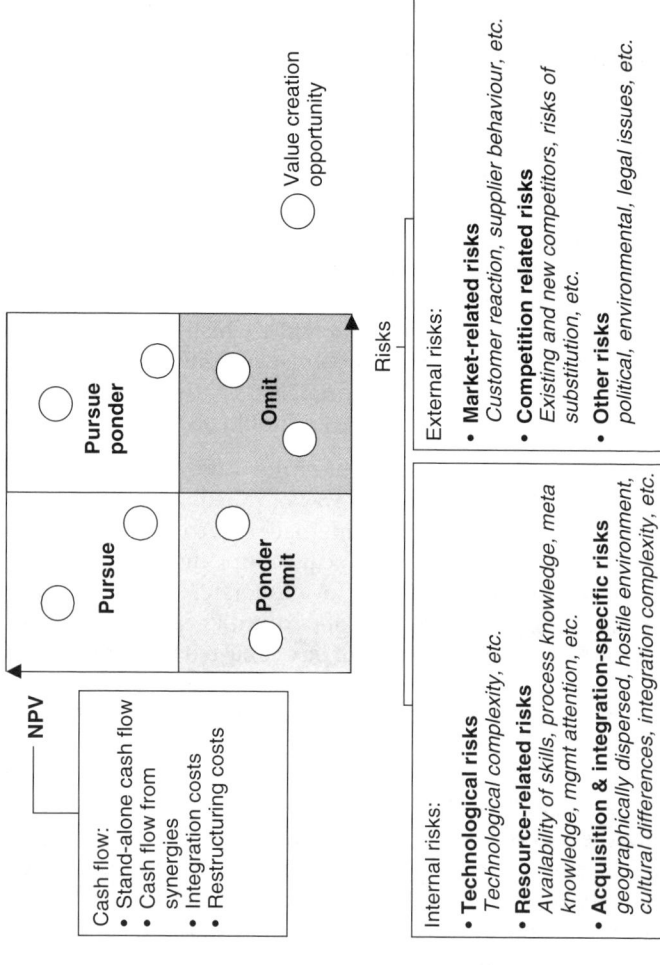

Figure 6.29 VCO portfolio and norm strategies

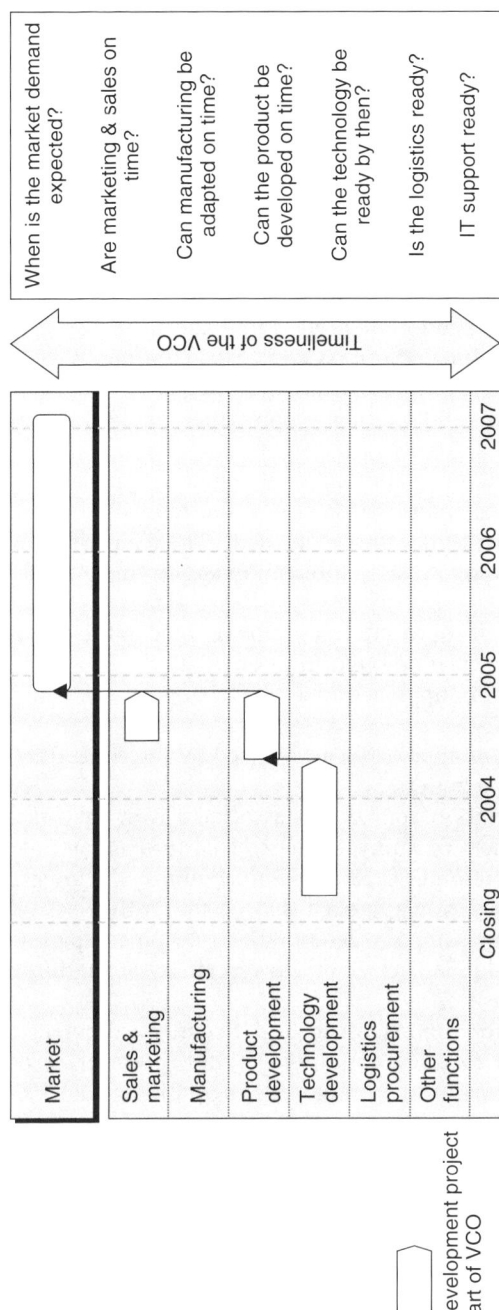

Figure 6.30 Timeliness of value-creation opportunity

was conducted on the functional and management layers. Now the question arises of which integration approach and company configuration is needed to achieve these strategic goals. Thus how shall the organizational structure, processes and behaviour of the merged companies look in order to enable the pursued value creation? Is it favourable to pursue a very close integration approach absorbing the target entirely or shall a slow and distant integration approach be in the fore? This early consideration of these integration aspects is critical to master the first phases of the integration process, which are mainly dominated by chaos, mistrust and the like. The consideration and determination of this integration strategy is not a separate task within the strategy-building phase, but needs to be addressed throughout the whole acquisition and integration process. The continuous responsibility and task to develop, revise and detail the integration strategy is held by the integration team and particularly by the involved transition manager (see Chapter 6.3.3.5).

The decision upon an appropriate integration approach, especially within innovation-driven industries, is one of the most difficult ones in strategic acquisition and integration management. The main challenges an acquirer faces when deciding upon an integration approach are the following:

- The integration approach needs to be adapted to the acquisition type and pursued value-creation opportunities;
- The integration approach may be changed over the course of the integration;
- The integration approach is influenced by the contextual characteristics of target and acquirer;
- A tight integration approach is associated with the risk of competence destruction, loss of innovativeness, departure of key inventors and specialists and a resource capacity overload during the integration;
- A distant integration approach is associated with a lack of value-creation potential, slow company deterioration, creeping departure of key inventors, high redundancies, 'ivory tower' effect, or the hindrance to emergent value creation;
- The integration approach can be different for different company areas and deferred to more appropriate times; however, it needs to finally amount to one holistic company configuration.

In order to appropriately consider all these challenges a procedure to determine an integration approach is proposed. It consists of two phases: (1) determination of the integration approach; (2) determination of the appropriate company configuration (see Figure 6.31).

This procedure can be applied to develop the overall company-integration approach or the configuration for each individual company area. Here the focus is laid on configuring the technology- and innovation-related areas of

Strategic Acquisition and Integration Management 191

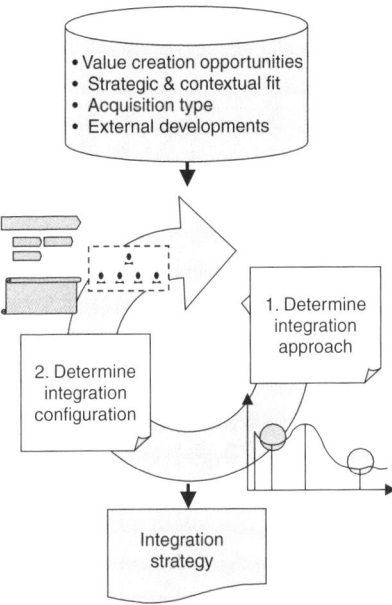

Figure 6.31 Developing integration strategy

the companies, enabling the achievement of technology-based value creation. The procedure is based on the model of the integration progress introduced in Chapter 5.4.2 and indicates how targets' and acquirers' organizational structures, processes and behaviour should be configured in the course of the integration and whether the integration should remain at a distance or proceed to a closer integration level. Figure 6.32 shows the relatedness between the pursued value-creation opportunities over time, the different integration phases and the required integration configuration and plan.

1. Determine integration approach In a first step it has to be decided whether the integration shall pursue only a distant approach or develop into a full and close integration. This is determined by the acquisition type and pursued value creation. The *distant integration approach* is characterized by its arm's-length organizational integration, its little overlaps in processes and in partly quite different company cultures. Its advantage is that the acquired target can continue its daily business as before and keep its innovativeness. The associated disadvantage of the low-integration approach is that joint value creation is difficult to pursue, company deterioration often happens, however slowly and unnoticed, and employee motivation decreases. Thus it is recommended that only a low-integration approach should be pursued if

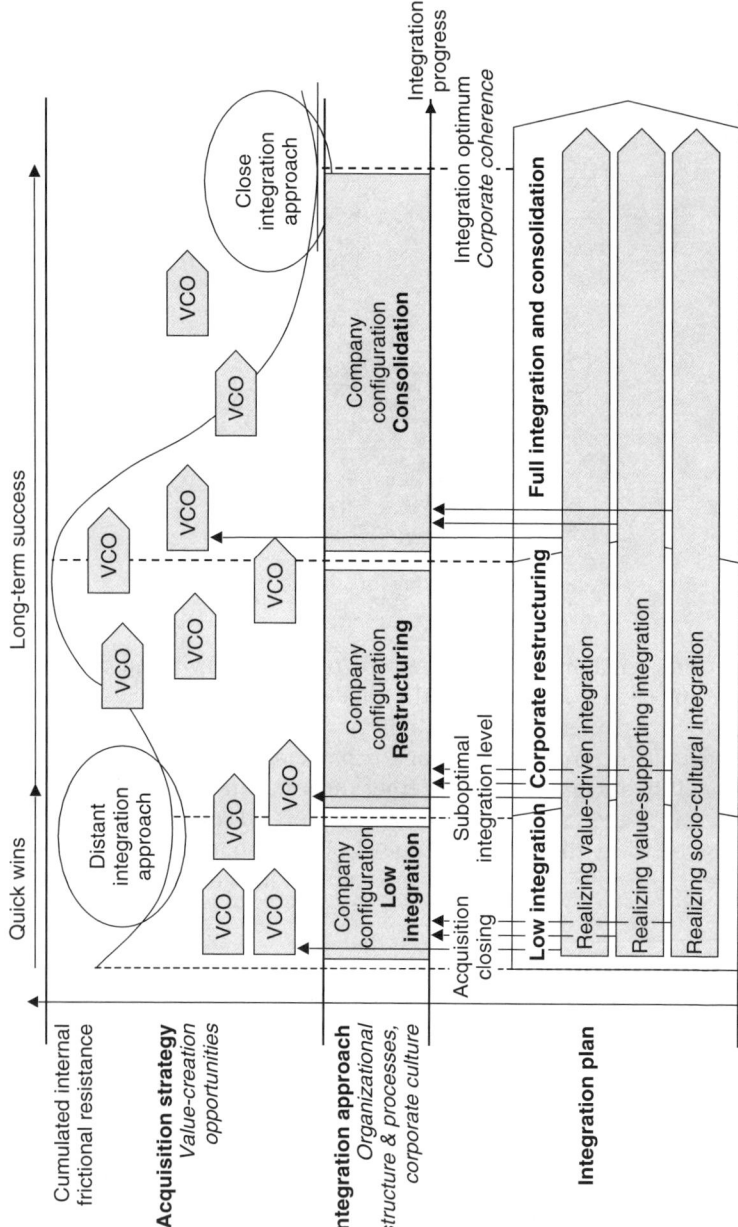

Figure 6.32 Integration process-related acquisition strategy, integration approach and plan

the acquisition focuses on value creation from quick gains such as cross-selling opportunities or value creation from leveraging easily combined elements or if the core technology of the target is in a pace-maker phase, when high innovativeness is absolutely crucial. Furthermore, non-related acquisition targets do not need to be integrated, although here the basic question can be posed why this target needs to be acquired anyhow if there is no or only a little potential value creation. Thus a low-integration approach is appropriate for only venturing acquisitions which do not aim for long-term knowledge transfer or resource deployment but focus on high and short-term innovativeness and the mere integration of products into their own product portfolio.

The *close-integration approach*, on the contrary, is suitable for long-term and especially cooperative value creation. The advantage of the close-integration approach is that the employees can be easily linked and thus the conditions for knowledge transfer and learning and, in turn, long-term success from innovation and resource deployment, are fulfilled. The disadvantage of a tight-integration approach is the risk of losing the key innovators, of reducing overall short-term innovativeness of the target and of facing cultural clashes. These risks have to be averted by the appropriate development of this close-integration phase. Thus narrow organizational integration, joint processes and an assimilated behaviour are conducive to long-term innovation and resource deployment if the target is very similar to the acquirer and medium- or large-sized. Figure 6.33 shows the attributes of those acquisition characteristics which determine the appropriate integration approach. Whereas some attributes demand a specific integration approach, others are more open-ended.

2. *Determining company configuration*   After having determined the appropriate integration approach, the concrete organizational processes, structures and behaviour and their change over time have to be derived. The appropriate considerations will be discussed under the headings of three integration phases (see also Figure 6.33) set out below.

- Company configuration for the low-integration phase: As mentioned earlier, the focus of the low-integration phase is to sustain the existing business and value creation and to reap some quick gains; cooperation and long-term value creation are not in the fore. This distant integration approach does not require tight structural or process integration. The companies are operated quite separately and only small and absolutely necessary integration efforts are pursued. The corporate cultures will also not highly assimilate or even integrate. The acculturation will rather result in separation or deculturation, which means either that the acquirer's and target's cultures drift away from each other or even build a negative, opposing attitude.[11] This deculturation often results in deteriorating performance of the target. Thus it is recommended that targets should be integrated at a distance only if they need to be managed as temporary internal ventures or the like.

- Company configuration for the restructuring phase: If the acquirer decides to pursue a tighter and close integration approach, it can either directly implement the according company configuration or first implement a temporary configuration which supports the restructuring of the two companies required for close integration. This detour, using temporary organizational structures, is especially useful in highly complex integration areas such as the R&D area.

Organizational restructuring needs to be enhanced via boundary-spanning structures, such as the exchange of individual employees or cross-divisional steering committees[12] or project groups. Furthermore, a temporary centralization of functions can be supportive to the restructuring effort.[13] Restructuring activities often result in the dissatisfaction of employees and their subsequent departure. Thus the organization has to be configured in such a way that key inventors or former owners of highly innovative companies can be retained and integrated within the overall corporate structure (see Best Practice 6

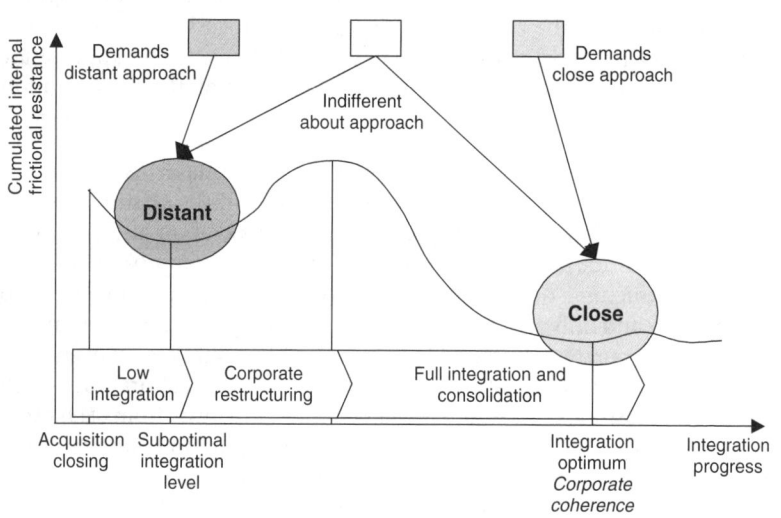

Figure 6.33 Factors determining the integration approach

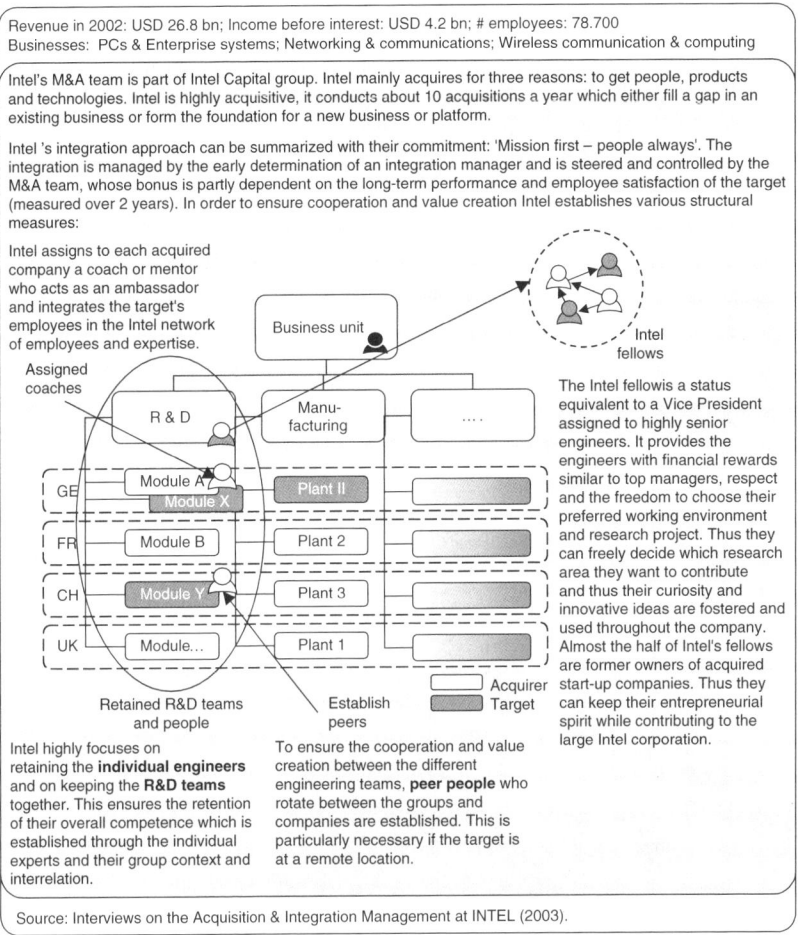

Best Practice 6 Structural integration measures at Intel

on Intel). Additionally, organizational measures can include the building of competence centres and the co-location of research groups. All these configural changes help to restructure the company and to enable the consolidation of the target and the acquirer. There are no particular processes to be implemented or cultures to be pursued as the restructuring phase is limited in time and thus does not demand further specific measures.

- Company configuration for the consolidation phase: The appropriate company configuration which fits with close integration is highly linked. However, within R&D attention has to be paid to not separating the engineering teams as their competence often lies in activities across the group (see Cisco in Best Practice 7). In accordance with the organizational structure the processes have to be redesigned and will need to cover both companies. A boundary-spanning innovation process, for example, ensures the inherent identification of synergistic technology-based value-creation opportunities. A framework for developing innovation processes and appropriate

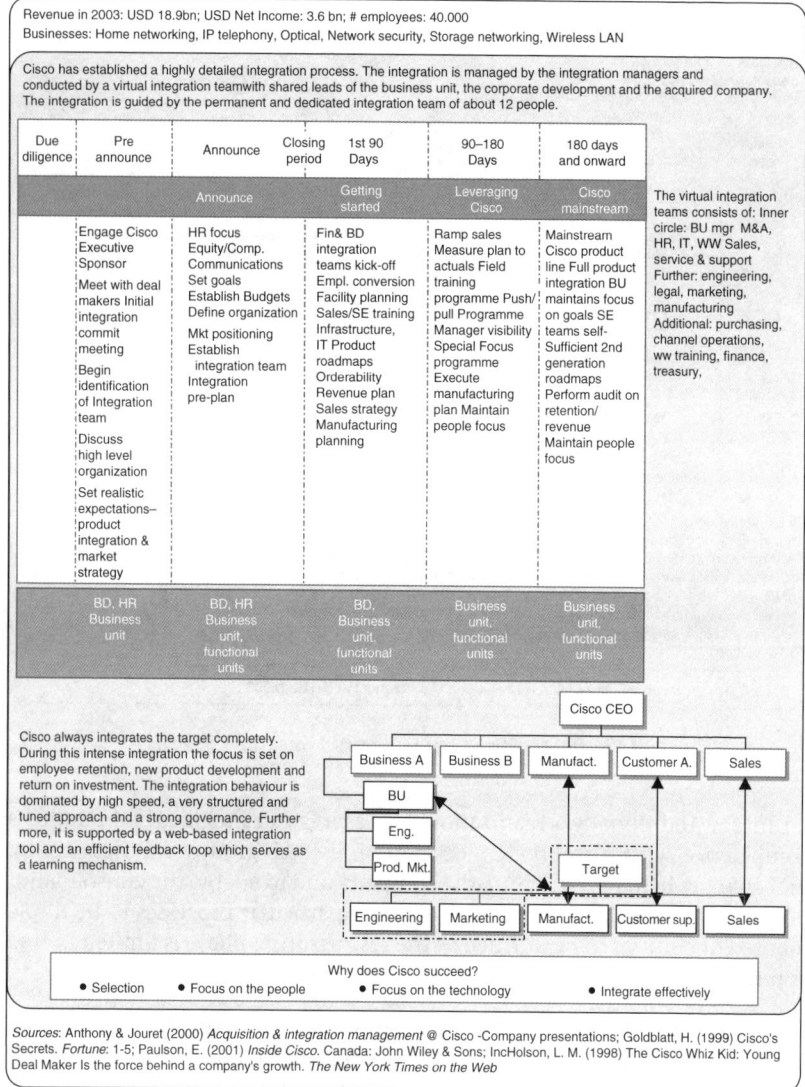

Best Practice 7 Integration management at Cisco

organizational structures was developed by Mitterdorfer-Schaad (2001). The behaviour that is sought in the close-integration approach is the integration of the different cultures or their assimilation.

As mentioned within the understanding of technology-based value creation, long-term innovation and resource deployment often emerge from the tight-integration approach. Thus the company configuration has to be tailored towards fostering the occurrence of emergent technology-based value-creation opportunities. One best practice of a technology-based value creation fostering company configuration can be observed at Ciba Specialty Chemicals (see Best Practice 8). Another aspect in determining the appropriate

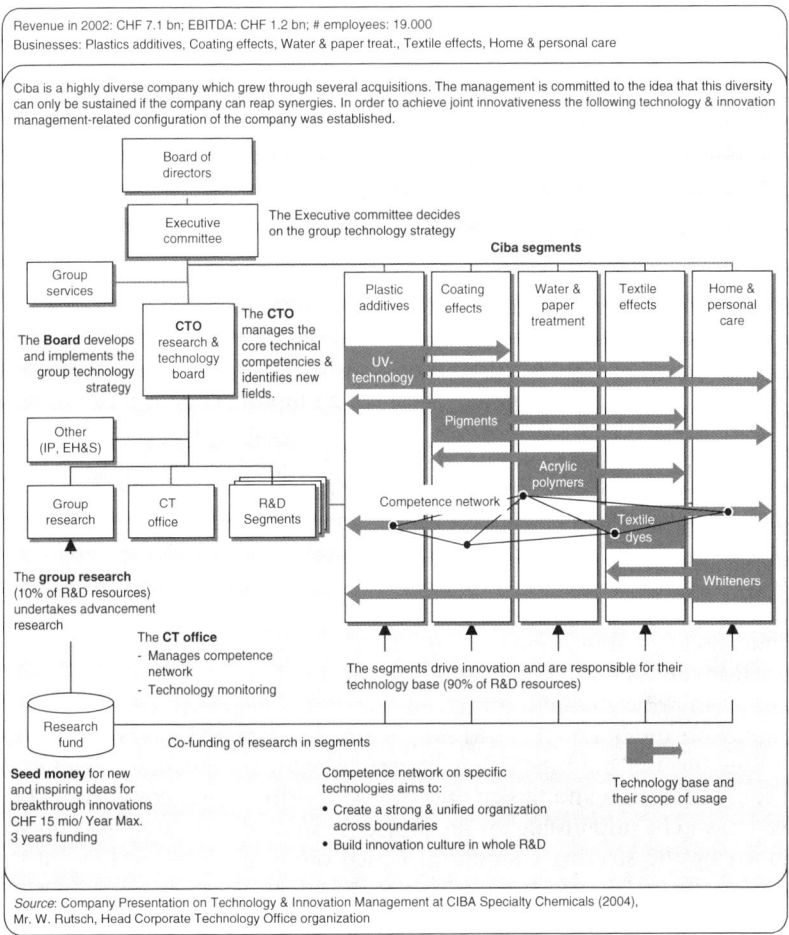

Best Practice 8 Technology-based value creation fostering company configuration at CIBA

company configuration for the tight-integration approach is the distinction between the normative, strategic and operational levels. Thus a tight and cooperation fostering company configuration has to be consistent on all levels. An example of a technology-based value creation fostering configuration on the normative level can be observed at Hewlett Packard (see Best Practice 9). This integration configuration has to be developed for all areas of the company and aggregated towards one holistic integration strategy.

The four steps of the acquisition and integration strategy-building process, company evaluation, identification of the value-creation opportunities and their assessment, as well as the determination of the integration approach and configuration, have to be summarized and aggregated from the different layers within the overall acquisition and integration strategy. This should include all relevant areas of the company and has to be similar to a business strategy of the merged companies. Furthermore, the financial planning of the merged companies based on the developed scenario of pursuing value-creation opportunities and following a specific integration approach can be developed. This holistic financial planning can be used to calculate the overall acquisition price of the target by considering synergy effects as well as integration costs accordingly.

6.3.3.5 Structural aspects

The question arises of who should be in charge of developing an acquisition and integration strategy. Here again the two different rounds of this phase have to be separated. Within the first preliminary round, with high information asymmetry and no direct target contact, the strategy development is conducted mostly by a corporate M&A team supported by the business unit, or if there is no corporate M&A team it is conducted by the business development. Within the second round, the structural aspects of the process become more concrete and split between the different layers. After the initial approval to pursue the business case and the contact to the potential target, a core acquisition team, a steering committee, functional and supporting teams as well as an integration-responsible team are established. The size and staffing of these groups depend on the size of the target. One representative acquisition structure is shown in Figure 6.34. The CTO should be part of the core team and should head the Technology and Innovation or R&D acquisition team which would consist of the head of engineering and R&D, production and IP experts, marketing personnel and product managers. This ensures the holistic integration of technology considerations within the overall acquisition and integration strategy. Furthermore, the business unit head has to be responsible for the project supported by his or her team and guided by the steering committee, which often includes the CEO of the holding, financial experts and the like. Additionally, one person from the M&A team, if it exists, should be part of the core team guiding the process. Furthermore, a transition manager has to become a member of the core

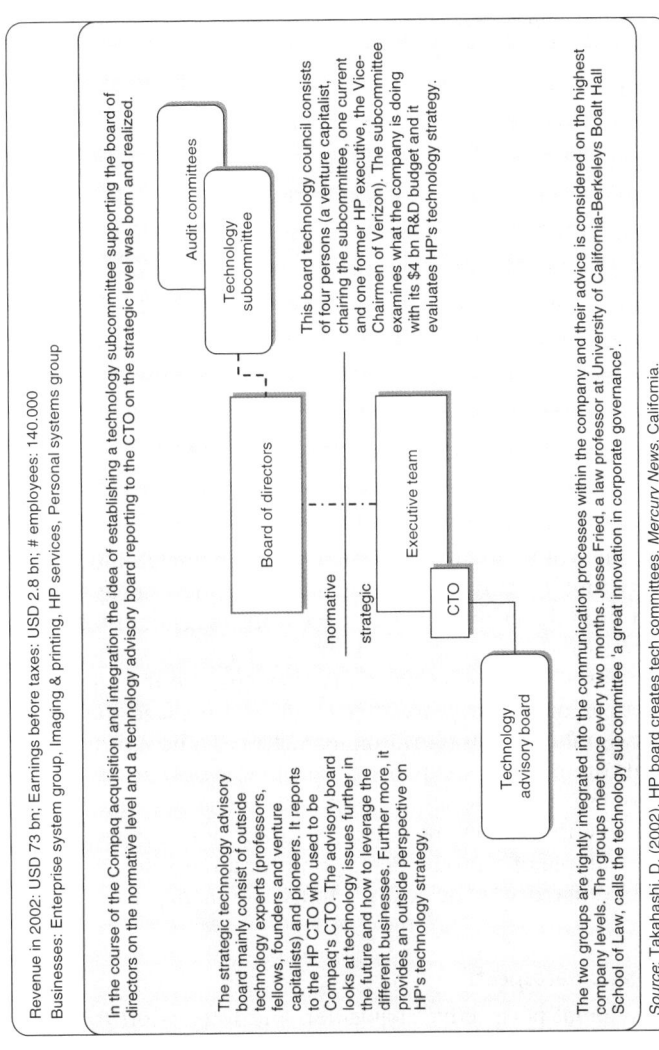

Best Practice 9 Innovation fostering context on the normative level at Hewlett Packard

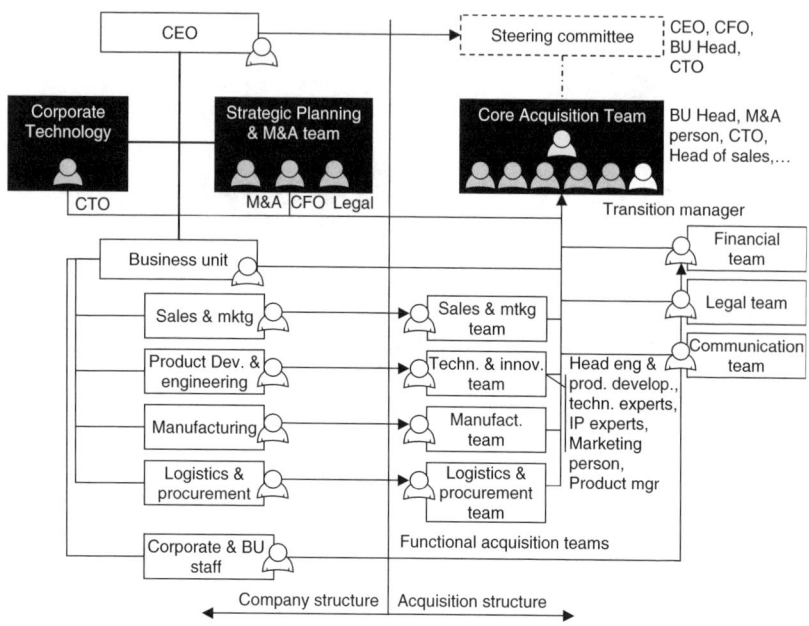

Figure 6.34 Example of an acquisition structure

team. He or she will be in charge of managing the integration process and thus it is very important that he be included into the strategy formulation process right from the beginning. He or she is in charge of bringing up integration concerns and of ensuring a long-term perspective on the acquisition. Therefore he or she needs to be 100 per cent assigned to the integration and usually has to remain two years after closing to manage the integration process. But who should this transition manager be? It became apparent that the technical competence or experience within one business is not the only decisive factor but rather the ability to act as an ambassador and leader at the same time is an important characteristic of a transition manager. Figure 6.35 provides an example of a profile description for a transition manager. One example of best practice for selecting a transition manager is IBM (see Best Practice 10).

6.3.3.6 Behavioural aspects

From a behavioural perspective acquisition and integration strategy development has to be managed in a diligent, friendly and open way. The acquisition team has to act from the understanding of an acquisition as a joint internalization of resources, an opportunity for corporate renewal for acquirer and target and as a means to achieve short- and long-term value creation. The target should be approached in a friendly manner indicating the levelled

TRANSITION MANAGER: PROFILE DESCRIPTION	CHECK
• Acts as an ambassador • Integrated in the whole process • Good leadership skills • Excellent communication skills • Good balance between pacing and trust building • Fair and open behaviour • Very good people networking • Not necessarily from the same business • 100 % assigned • Reward of management, business development & transition manager depending on integration success • Reporting to an integration steering committee • Tightly linked to management	

Figure 6.35 Profile description of a transition manager

In order to become a Vice President (VP) of IBM you have to fulfil two conditions:
• To have worked and lived abroad
• To have been an integration manager of an acquisition.

IBM has a worldwide database of potential transition managers, who also have the potential to become a VP. Thus the transition manager does not necessarily come from the acquiring business.

This policy ensures that the decision-makers, the Vice Presidents and above, have themselves experienced the challenges of a target integration and thus will diligently consider the long-term effects of an acquisition.

Best Practice 10 Transition managers at IBM

relative standing between the companies. Furthermore, the acquisition team has to learn to manage risks and information asymmetries. An acquisition and integration strategy is always based on assumptions and thus will not reflect absolute truth. Instead of eagerly trying to reduce the information asymmetry and risk direct conflict with the target resulting in a locked-in situation, the acquisition team has to manage risks, and show that by prioritizing and reducing the risks they were both necessary and bearable. Additionally, the core team has to make sure that it does not get trapped within the accelerating cycle of the acquisition process until a point of

> In order to ensure the diligent acquisition behaviour and the target's fit to CIBA Specialty Chemicals, the company's CTO has established an acquisition team which has the task to challenge current acquisition projects, to point at potential up coming risks and to question the target's fit to CIBA. This committee is regularly consulted and informed. It consists of several highly experienced and long-term CIBA employees all from different areas. The CTO and the head of the Corporate Technology Office are also part of this 'challenging committee'.

Best Practice 11   M&A Challenging Committee at CIBA

no return. Thus this accelerating momentum has to be actively reduced. An example of best practice in this field is seen at Ciba Specialty Chemicals (see Best Practice 11).

6.3.3.7   Conclusion

Thus it can be concluded that the development of the acquisition and integration strategy consists of four phases: the analysis of the target; the identification of value-creation opportunities; their assessment; and determination of the integration approach and configuration. These activities are conducted in two rounds, reducing information asymmetry between target and acquirer. The findings are summarized within a preliminary business case which contains the strategic objectives of the acquisition, the appropriate integration approach and the financial model with a stand-alone target value and an acquisition price including synergy effects and integration costs.

This proposed procedure for developing an acquisition and integration strategy ensures that the process is not only pursued from a short-term and financial perspective but that technology-based value-creation opportunities are considered early enough, that the process is managed in an integrative, systemic and holistic way, and that the increasing momentum can be mastered.

6.3.4   Due diligence

The developed acquisition and integration strategy, which is nothing more than a scenario based on assumptions, needs to be approved by the board in order for the Letter of Intent[14] to be signed. This legally binding agreement officially marks the intention of the acquirer to buy the target and is the initiation of the Due diligence phase which is used to reduce the remaining information asymmetry between target and acquirer. *The due diligence* aims to *validate the acquisition and integration strategy*, to *identify risks* and to *support the price finding*. The due diligence is separated into different areas at the different layers, such as legal, financial, tax, strategy and market, technology and production and environmental due diligence. Each due diligence is conducted in a different way: whereas those for the legal, financial and tax are mostly conducted by external experts in separate data rooms and have a

Strategic Acquisition and Integration Management

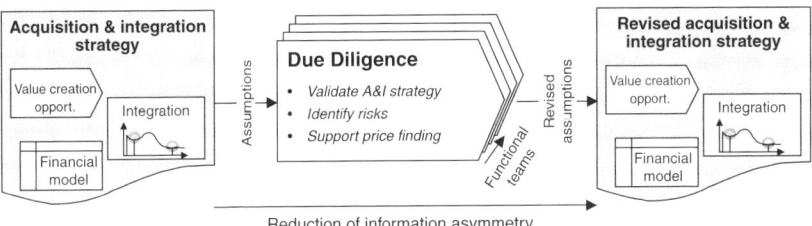

Figure 6.36 Due diligence

very limited duration, the due diligence on the functional layers, such as the strategy and market, technology-related and environmental due diligence, are not so separated. They run in parallel with the strategy development phase and are conducted via direct interviews, site visits and also the investigation of documents. The findings of all due diligence areas are used to revise the acquisition and integration strategy in order to arrive at a holistic document which serves as the basis for deciding upon signing the acquisition or not (see Figure 6.36).

Whereas there is a vast amount of literature[15] discussing the critical aspects of a legal, tax, financial and even environmental due diligence, the technology due diligence has so far received only very limited attention. Thus while this chapter focuses on the tasks and methods of the technology due diligence, the considerations actually partly overlap with the market and strategic due diligence. Furthermore, the production or technical due diligence as well as the innovation considerations are also integral elements of the overall technology due diligence. The overall checklist to apply during a technology due diligence can be found in Appendix B.

The elaboration of the understanding of technology-based value creation in corporate acquisitions has shown which aspects need to be considered within the technology due diligence in order to ensure future technology-based value creation:

- Ensure the technology-based value creation potential and thus the future possibility for realizing planned and emergent technology-based value-creation opportunities;
- Ensure that the external developments are appropriately considered;
- Ensure the contextual fit between the target and acquirer and thus the ability to successfully realize the value creation potential;
- Validate the technology-based value-creation opportunities in a prospective way.

The technology due diligence validates the technology-based value creation potential and the ability to realize it. Thus it aims to ensure the occurrence and profitability of planned and emergent technology-based value-creation

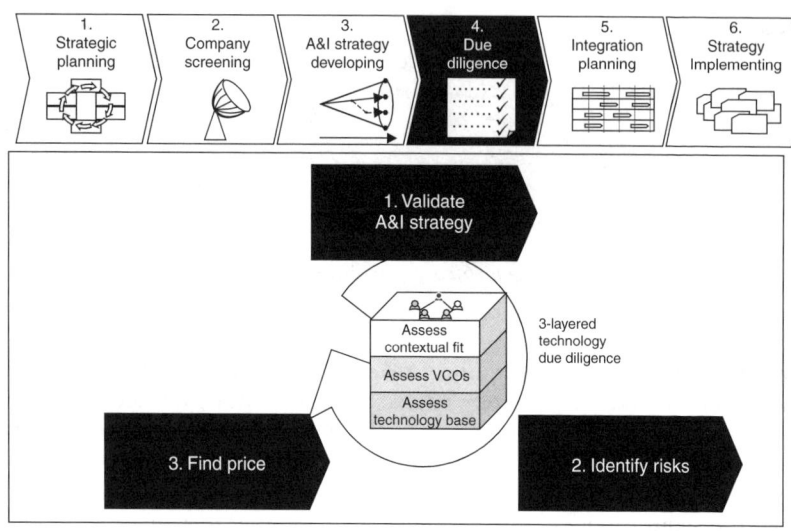

Figure 6.37 Technology due diligence

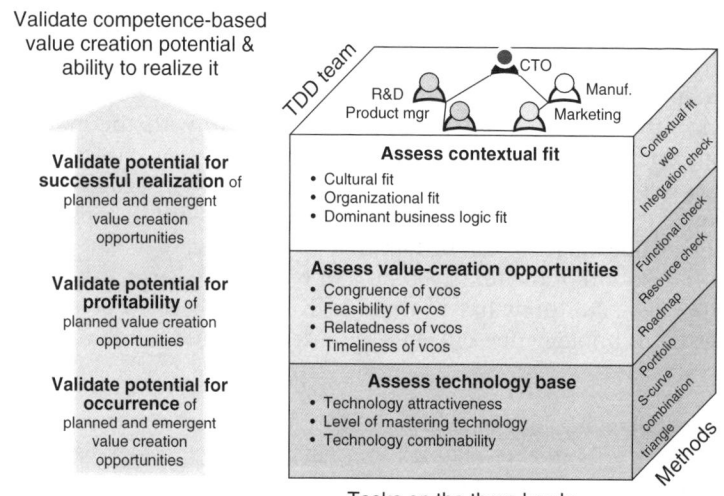

Figure 6.38 Three levels of the technology due diligence

opportunities and the joint capability to realize these. This validation takes place on three different levels of investigation. Each level has another object of investigation and addresses all three tasks: validation of the A&I strategy; identification of risks; and support of price finding (see also Figure 6.37). Figure 6.38 provides an overview of the three levels of the technology due

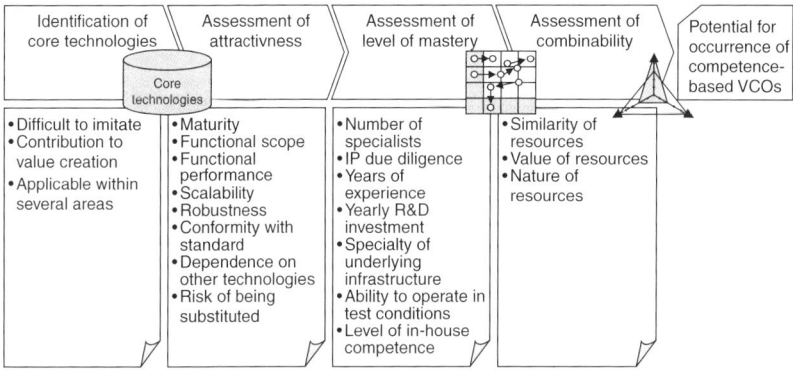

Figure 6.39 Process of the first-level technology due diligence

diligence and the associated tasks, methods and structural aspects which will be explained in the following section.

6.3.4.1 First-level technology due diligence – assessment of the technology base

The first level of the technology due diligence validates the occurrence of planned and also emergent technology-based value-creation opportunities (see Figure 6.39). This occurrence is determined by the technology base of the target. Thus an attractive, highly mastered and particularly combinable technology base fosters the realization of planned value-creation opportunities and the future emergent integration of resources into innovation and resource-deployment opportunities. The question arises how these characteristics of the technology base can be assessed within the technology due diligence.

First of all the investigation cannot cover all technologies within the company. The assessments have to focus on the core technologies which have the highest potential to be integrated as a crucial resource for value-creation opportunities. The core technologies can be identified by the difficulty of imitating them, by their existing or potential significant contribution to the target's value creation and by their potential application within several areas of the company. It is important that the investigation addresses not only existing core competencies but also potential ones which are still in development.

As a next step the attractiveness of the different core technologies has to be assessed. Attractiveness can be measured by a variety of attributes. The following criteria are a summary of various indicators for technology attractiveness:

- Maturity (pace-maker, key or base technology)
- Functional scope (discontinuous, wide or narrow)
- Functional performance (in relation to substitutes)

- Scalability
- Robustness
- Conformity with standard
- Dependence on other proprietary or costly product or process technologies
- Risk of being substituted.

This attractiveness is an indication of whether the technologies are mature and attractive enough for integration into joint product developments or should be sold into new markets. Furthermore, an attractive technology base fosters resource transfer and integration and thus the emergence of hidden technology-based value-creation opportunities. Technological attractiveness cannot be measured within a data room. The prototypes and first machines delivered to the customer site under operational conditions need to be visited and observed. It is important to take the opportunity to eventually even test and operate the machine to get a feeling of how sound the attractiveness of the underlying technologies actually is. In the case where little technological attractiveness is found due to proprietary or immature technologies, the business case needs to be revised as the technological problems or immaturities of the company need to be solved before realizing the synergies. Thus a low technological attractiveness bears the risk of additional development costs and long delays.

In a next step the *level of mastering the technology base* has to be investigated. The most attractive technology will not contribute to value creation if it is not mastered. The level of mastery of a technology can be measured by the following factors:

- Number of designated specialists within the field
- Number, soundness and quality of patents within this field (IP due diligence)
- Years of experience
- Yearly R&D investment in the field
- Specialty of underlying infrastructure
- Ability to operate in test conditions
- Level of in-house competence versus external abilities.

The assessment of the level of mastery is a strong indication of where the real competencies of the target lie and which resources should be the base for value creation and which should not. A high level of mastery motivates people to cooperate and fosters the knowledge flow. The lack of mastery, to the contrary, can result in risks such as additional financial and timely efforts to increase the level of mastery or an inability to develop and launch a joint product due to a lack of skills or patent problems.

The analysis of the level of mastering the technology base has to cover the target's core technologies as well as the ones from tightly linked suppliers. As these are often tied to the target via long-term contracts their technological ability also has an important impact on the resulting value creation potential.[16] Furthermore, the assessment should address not only the current level of mastery but also the pursued future focus of the target. This is a strong indication of where technological competence sits. The analysis needs to integrate the opinions of suppliers, competitors and customers, if this is possible, in order to obtain a good overview and an intersubjective estimation of the target's level of mastering its technologies. Other highly important aspects for assessing of the level of mastering the technology base are an IP due diligence and a site visit which investigates the maintenance and specialty of the used infrastructure. There is some literature[17] on IP due diligence and a best practice example is provided by Unaxis (see Best Practice 12).

The attractiveness and the level of mastering the technology base over time can be visualized within the dynamic technology portfolio (see Figure 6.40). It provides a good overview of the technologies of the target and indicates which resources will foster the emergence and realization of technology-based value-creation opportunities and which will not contribute to value creation in the short or medium term. These findings can be

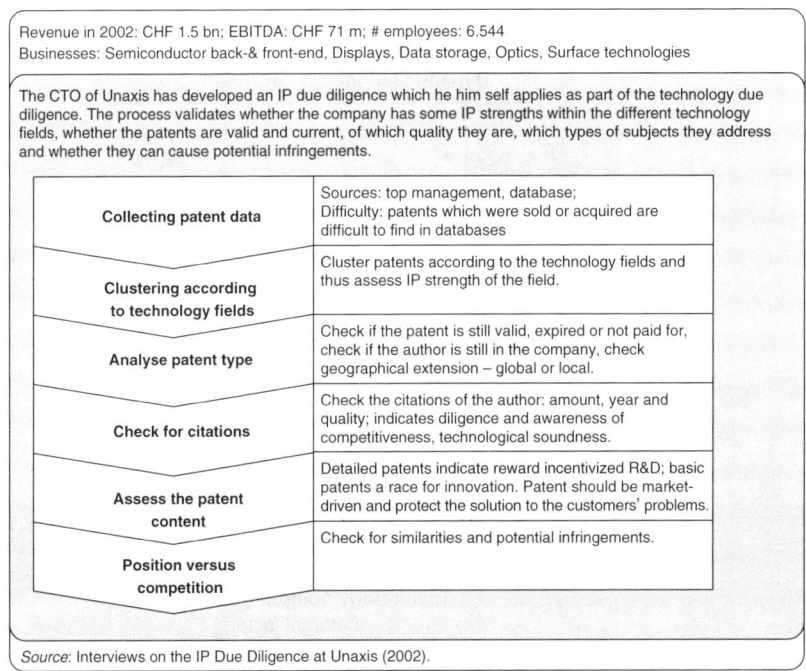

Best Practice 12   Intellectual property due diligence at Unaxis

used to revise the ideas of value-creation opportunities if, for example, the technology is neither attractive enough nor well mastered.

The last step in the validation of the potential for occurrence of planned and emergent value-creation opportunities addresses the *combinability of the technologies and resources*. It was observed that technology-based value-creation opportunities occur particularly where the technology bases can be combined. Combinability of a resource depends on the pursued value creation. The transfer or substitution of resources, such as the transfer of CAD or manufacturing technologies, or quick gains, such as the leveraging of a module, require highly similar or complementary resources which can be easily transferred and are thus not very context-dependent and are neither dispersed nor particularly explicit. On the contrary, the long-term integration and learning of competencies which aim to create highly attractive new innovations need the integration of highly rare, unique and supplementary technologies. Thus Figure 6.41 shows that the characteristics of an appropriate combinability depend on the pursued value creation or acquisition type. Thus at 'Venturing Acquisitions' the target's technology base should be highly valuable and supplementary or distinct to the acquirer's resources. This, however, also indicates that these resources will be very difficult to integrate in the long run due to their tacitness, context-dependence and dispersion. At 'substrate-for-growth' or 'play-for-scale' acquisitions the target's technology

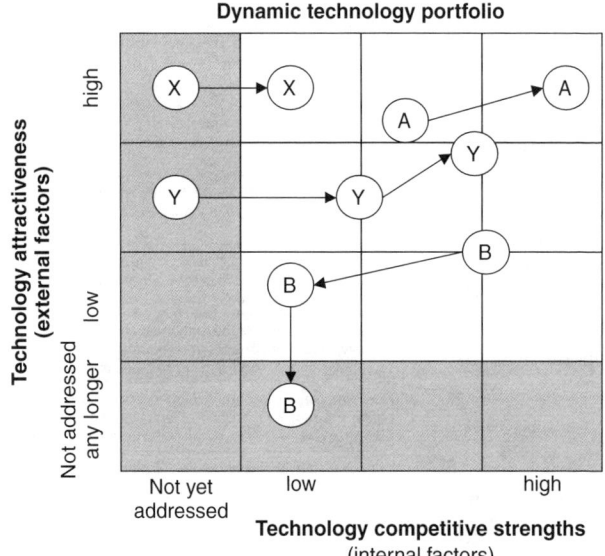

Figure 6.40 Dynamic technology portfolio
Source: Tschirky (1998c: 313).

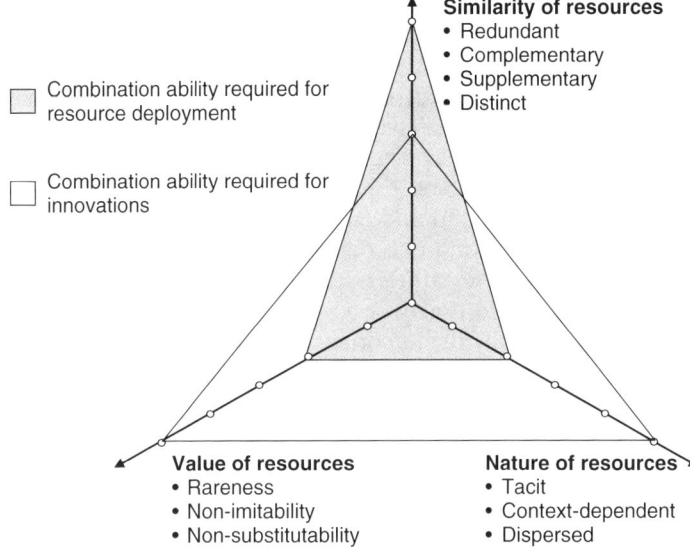

Figure 6.41 Required level of combinability depending on required VCO

base should rather be more easy to integrate and of a complementary nature. These considerations have to be validated during this phase of the technology due diligence. The following questions can be posed:

- Are the technologies sufficiently similar/redundant or complementary/ supplementary/distinct?
- Is it possible to integrate the technologies in the long run due to their tacitness, context-dependence, dispersion or e.g. non-modular structure?
- Are the technologies too specific, rare, or difficult to imitate to transfer them?

After these investigations, the technology due diligence team can assess the potential for the occurrence of planned and future emergent value-creation opportunities. If the technology base is well mastered and attractive and can be combined according to the intentions then the potential for short- and particularly long-term technology-based value creation of the merged companies exists.

6.3.4.2 Second-level technology due diligence – assessment of value-creation opportunities

The second level of the technology due diligence validates the attractiveness and profitability of the identified and planned value-creation opportunities and thus of the joint product and technology pipeline. It needs to be verified that the identified highly strategic and planned value-creation opportunities are as profitable, timely and associated with a certain risk level as initially assumed. These investigations are conducted by assessing the congruence, relatedness and feasibility of the value-creation opportunities and by revising their timely occurrence (see Figure 6.42). These attributes of a value creation opportunity determine whether the project will become a financial success. The analysis is tailored towards innovation projects and is thus applicable to both innovation and resource deployment projects, whereas the first is a clear innovation project and the second can only indirectly be linked to an innovation. Nevertheless, the investigation on the congruency, relatedness, feasibility and timely occurrence is appropriate for both value creation mechanisms; only the importance of the different aspects varies.

In a first step, the *congruence of the planned value-creation opportunities* is assessed with the following criteria. A technology-based value-creation opportunities must:

- Represent a missing link between customer needs expressed in functional requirements and functional scope provided by the linked technologies;
- Balance the market pull and technology push;
- Have a modular, complexity-reducing and elegant basic underlying (product) concept.

This congruence of the value creation opportunity ensures uniqueness to the customer and a sufficient value contribution. The modular and sound underlying concept provides the basis for being able to adapt the value propositions to customers' need in a flexible way and to keep the costs of realizing a new product or service in its barriers.

In a second step the *relatedness of the pursued value-creation opportunities* to the existing business is evaluated. It was observed that joint development projects of a completely new technology or product are highly difficult to achieve, especially as there is insufficient experience within this new area. Thus, especially for the first value creation activities, it is appropriate to stick to the field of experience. The lack of this relatedness is also an explanation of the difficulty of mastering unrelated acquisitions. Relatedness can be assessed by asking the following questions:

- Are we acquainted with the customers and markets which we address with the value-creation opportunity?
- Have we ever managed a similar product concept before?
- Do we have sufficient experience within the technology platform?
- Is the way to develop or service the product similar to our way of doing business?

The congruence and relatedness assessment can be well conducted using the innovation architecture (see Figure 6.43).

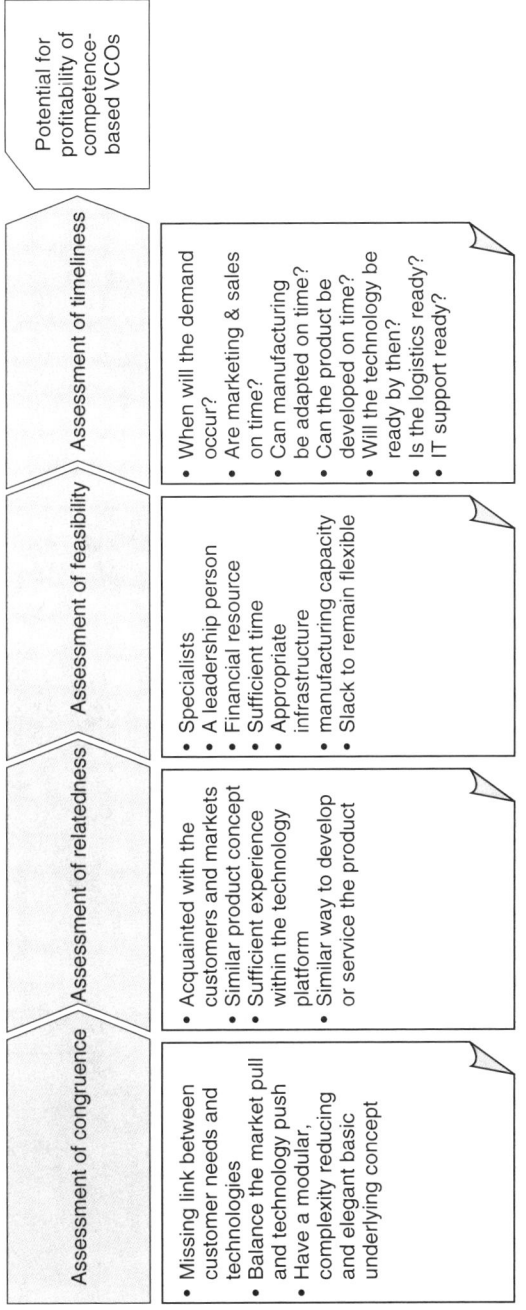

Figure 6.42 Process of the second-level technology due diligence

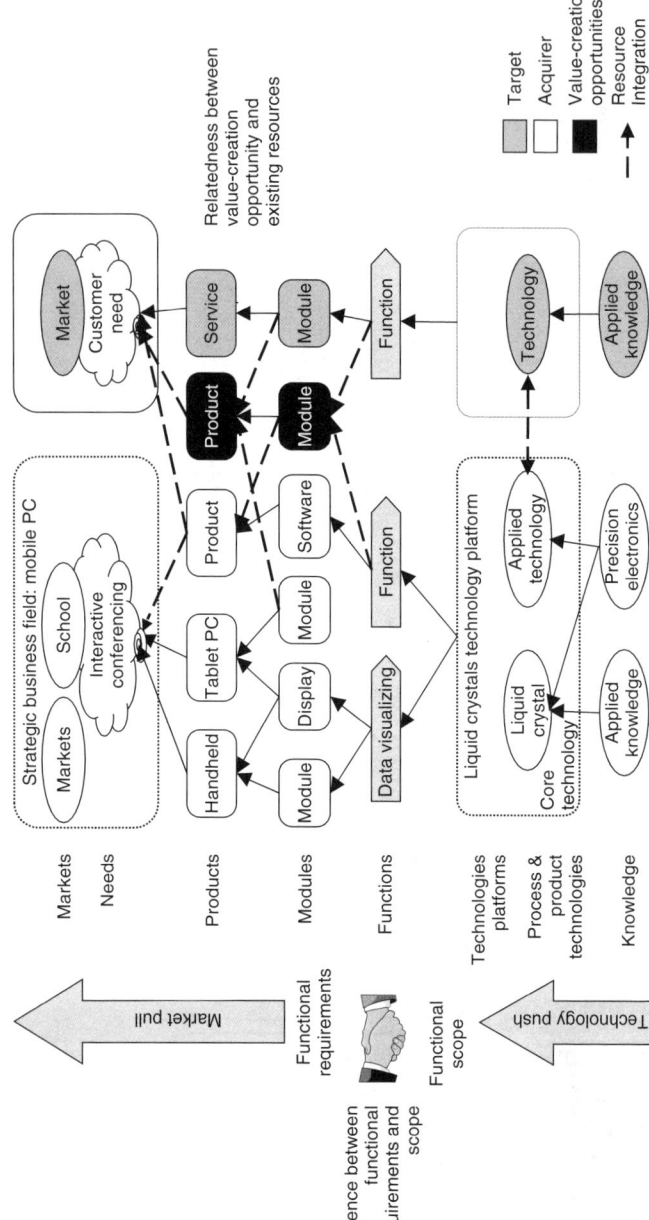

Figure 6.43 Assessment of congruence and relatedness using innovation architecture

The third investigation, validating the potential for profitability of the value-creation opportunity, is the *feasibility check* which is concerned with the availability of resources to develop and manufacture the value-creation opportunities. The development of new technologies or products or the transfer of knowledge requires the existence and availability of certain resources. These resources cover the object knowledge about the resource, the methodological knowledge on how to transfer, develop or integrate the resources and a certain amount of meta-knowledge which ensures that the integration of the known resources can happen within the context. Additionally, the realization of the developed product or the operation of the transferred manufacturing machines requires sufficient manufacturing skills and particular capacities. The main topics to address are the sufficient availability and assignment of:

- Specialists, people knowledgeable within the area
- A leadership person managing the realization of the value creation opportunity
- Sufficient financial resources
- Sufficient time
- The appropriate infrastructure
- The required manufacturing capacity
- Sufficient slack to remain flexible.

The last assessment of the profitability of the identified value-creation opportunities is the revision and validation of its timeliness. In Figure 6.30 the assessment of the timeliness of value-creation opportunities as part of their assessment within the acquisition and integration strategy formulation was shown. This assessment has to be revised with the collected data from the internal capabilities and resources and external market data inferred from the market due diligence.

The result of the validation of the profitability of the originally identified value-creation opportunities is used to revise their associated NPV, risk level and timeliness. Thus a more realistic profitability of the value creation potential can be assessed which again supports the price finding. Additionally, a lack of congruence, relatedness, feasibility and timeliness points to upcoming risks which have to be marked by the flag methodology, compensated for, eliminated or even averted by not continuing the acquisition.

6.3.4.3 Third-level technology due diligence – assessment of contextual fit

The third level of the technology due diligence validates the *ability to finally realize the planned and emergent value-creation opportunities* (see Figure 6.44). This is highly determined by the contextual characteristics and similarities of the target and the acquirer. Thus characteristics of the target's corporate context, such as of the organizational structure, corporate culture or dominant

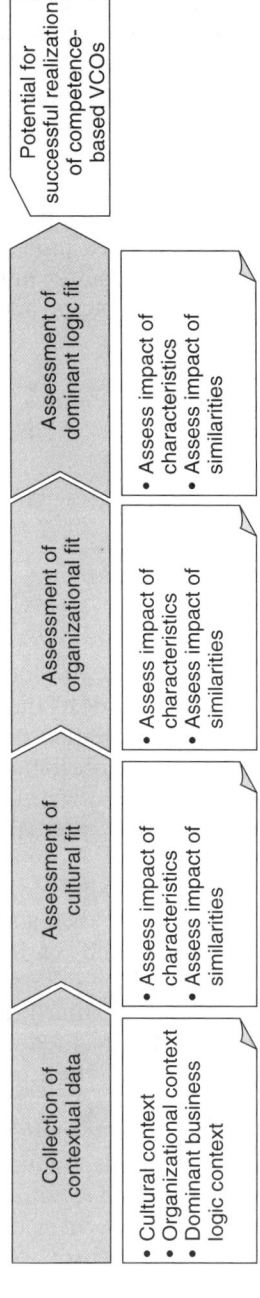

Figure 6.44 Process of the third-level technology due diligence

business logic including the top management's abilities, which are fostering the innovativeness and efficient redeployment of resources and are quite similar to the acquirer's context, increase the probability of successfully realizing the planned and emergent value-creation opportunities. Even though the contextual fit was already addressed within the screening phase, within this phase the level of detail can be increased and the direct contact with the target during the due diligence supports the investigation.

In a first step the characteristics of the target's corporate context need to be collected, given the knowledge of its own contextual attributes. The information concerns the corporate culture, the organizational characteristics and the dominant business logic of target and acquirer. This collection and the subsequent assessment are not the task of one individual person. These types of soft information need to be gathered by all members of the acquisition team throughout the whole process and aggregated and discussed within joint meetings. Thus each person involved throughout the whole acquisition process and particularly during the technology due diligence process should actively observe the contextual characteristics and consider their implications. Figure 6.45 shows an example of the main contextual characteristics within the fields of cultural, organizational and dominant business logic context which need to be investigated. The specific attributes were introduced earlier in Figure 6.19. For example, a highly entrepreneurial leadership within the R&D area is conducive to technology-based value creation, whereas a low connectedness between different R&D sites hinders cooperation and thus joint innovation. The different characteristics can be visualized in a contextual web (see Figure 6.45).

Once the data are collected, the assessment of the contextual fit is initiated. First the *fit of the corporate cultures* needs to be assessed. As outlined in Chapter 5.3.4.cultural fit is determined by the conduciveness to innovativeness of target's and acquirer's cultural characteristics and by their similarity. Whereas a quite high level of similarity needs to be pursued, too close similarity might cause a lack of sense of urgency and curiosity and is thus also not favourable (see Figures 5.13 and 6.46). Thus the following questions need to be addressed when assessing the cultural fit that will be conducive to technology-based value creation:

- Are both companies' openness conducive to technology-based value creation and sufficiently similar?
- Are both companies' leadership styles conducive to technology-based value creation and sufficiently similar?
- Are both companies' time managements conducive to technology-based value creation and sufficiently similar?
- Are both companies' reactions to change conducive to technology-based value creation and sufficiently similar?
- Are both companies' diversities conducive to technology-based value creation and sufficiently similar?

216 Mastering the Acquirer's Innovation Dilemma

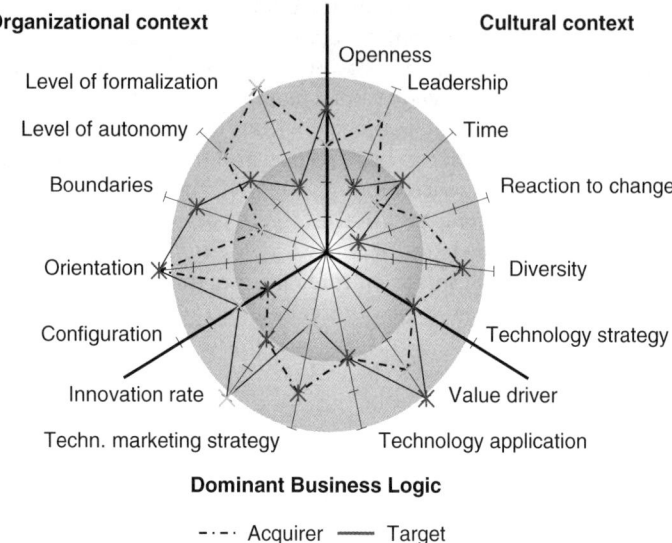

Figure 6.45 Contextual web of the target's and acquirer's characteristics

The procedure within the assessment of the organizational and dominant business logic fits accordingly. The only difference is that some contextual differences within the cultural and dominant business logic contexts are favourable, whereas the organizational fit is best achieved if the contexts are very similar.

The measurement of fit can be conducted with quantitative measures (see Figure 6.46). The sum of innovation-fostering contexts indicates whether the characteristics of the corporate contexts are conducive to technology-based value creation, and the absolute value of the difference[18] between the characteristics of the corporate contexts are an indicator of whether the similarity between the contexts is conducive to technology-based value creation.

An example of how the calculation of contextual fit can be conducted is given in Figure 6.47. The evaluation of the contextual fit indicates whether the two companies will be able to closely work together and to push innovativeness and help to estimate upcoming integration efforts and problems. Furthermore, it identifies potential for improvements for both acquirer and target and upcoming risks such as organizational clashes or cultural conflicts. The findings should be used to revise the planning of value-creation opportunities, integration approach and also integration efforts considered within the price finding.

Even though contextual differences are often not considered as a cause for discontinuing an acquisition, nonetheless the findings should raise the

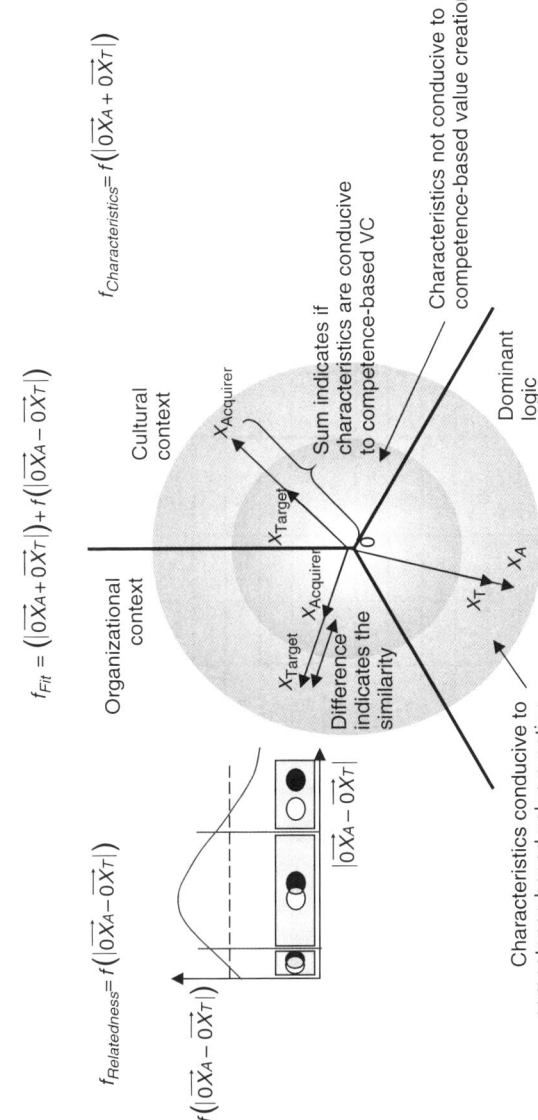

Figure 6.46 Calculation of contextual fit

218

Context	Characteristic	Acquirer	Target	Characteristics conducive to cbvc		Relatedness conducive to cbvc			Total: Weight 1: 0.8	
				Sum	Score	Dif.	Score		Score	
Cultural Context	Openness	3	4	7	4 good	1	5 very good	72%	good	
	Leadership	4	2	6	3 medium	2	3 medium	50%	medium	
	Time	2	3	5	3 medium	1	5 very good	61%	good	
	Reaction to change	3	1	4	2 poor	2	3 medium	39%	poor	
Dominant logic context	Diversity	4	4	8	4 good	0	4 good	67%	good	
	Technology strategy	3	3	6	3 medium	0	4 good	56%	medium	
	Value driver	4	5	9	5 very good	1	5 very good	83%	very good	
	Technology application	3	3	6	3 medium	0	4 good	56%	medium	
	Technology marketing strategy	2	4	6	3 medium	2	3 medium	50%	medium	
Organizational context	Innovation rate	5	3	8	4 good	2	3 medium	61%	good	
	Configuration	3	2	5	3 medium	1	5 very good	61%	good	
	Orientation	5	5	10	5 very good	0	4 good	78%	very good	
	Boundaries	2	4	6	3 medium	2	3 medium	50%	medium	
	Level of autonomy	4	3	7	4 good	1	5 very good	72%	good	
	Level of formalization	5	2	7	4 good	3	2 poor	56%	medium	

9–10	5	very good	4 good	81%–100% very good
7–8	4	good	5 very good	61%–80% good
5–6	3	medium	3 medium	41%–60% medium
3–4	2	poor	2 poor	21%–40 % poor
1–2	1	very poor	1 very poor	1–20% very poor

Figure 6.47 Example of the calculation of contextual fit

acquirer's awareness of the upcoming challenges and risks. Furthermore, they will be integrated with the findings from the other due diligence areas and layers to adapt the holistic acquisition and integration strategy, which then serves as a decision-making base for either continuing or discontinuing the acquisition.

6.3.4.4 Structural and behavioural aspects

The question arises of how the technology due diligence should be conducted and especially who should be in charge of it. Generally the technology due diligence should be the task of the technology and innovation functional subteam of the acquisition. This team should be staffed with technical experts within the research, engineering and manufacturing, product managers, marketing experts and IP specialists. It should be headed by the CTO or the head of engineering of the company. Many companies hire external technological experts to support the technology due diligence. On the one hand this approach is legitimate if the acquirer has insufficient knowledge of the specific technology, on the other hand it is recommended that the acquirer first builds up some internal competence or absorptive capacity to be able to investigate the technology and subsequently to internalize and transfer it. An external expert neither knows what the acquirer intends to do with the technology in combination with its strengths nor can he or she help to internalize the technology in question. Thus while external expertise is suitable for complementing the technology due diligence team it should not be the only experienced unit within that field.

From a behavioural point of view the technology due diligence should be dominated by risk awareness, a focus on future developments and potentials and sufficient diligence and patience in order not to get trapped in the rush of the accelerating momentum.

6.3.4.5 Conclusion

It can be concluded that the technology due diligence aims to reduce the information asymmetry between target and acquirer, validate the acquisition and integration strategy, identify risks and support the price finding. This is achieved on three different levels. On the first level the potential for occurrence of technology-based value-creation opportunities is validated by assessing the target's technology base. On the second level the potential profitability of the identified strategic technology-based value-creation opportunities is validated by assessing their attractiveness and on the third level the ability to successfully realize the technology-based value creation potential is verified by assessing the contextual fit. The investigation by the technology due diligence team needs to be conducted in close cooperation with the target company and its stakeholders to follow a prospective perspective and to be integrated within the overall A&I strategy revision. The advantage of the approach is that it ensures the potential for long-term and also emergent value creation and is not only focused on the existing capabilities. Furthermore, the proposed methods and the specific focus of each layer ensure the applicability of the concept. This new approach prevents acquirers from buying immature technologies or low competencies, from underestimating technological risks and from paying an exaggerated acquisition price.

6.3.5 Integration planning

So far it has been determined which value-creation opportunities can be achieved with the target, which integration approach and configuration are appropriate to realize these value-creation opportunities and how much should be paid for the target. Now the focus has to shift from strategy formulation towards the strategy implementation. Thus the fifth phase of the acquisition process, which actually also runs in parallel with the strategy formulation and due diligence process and can last until the closing of the acquisition, addresses integration planning. This integration planning aims to determine which actions are required to achieve the acquisition and integration strategy and how and when these should be pursued. Thus integration planning directs all actions after the acquisition which concern value creation-related projects and the accompanying change in company configuration.

Integration planning is managed and guided by the transition manager on the integration layer and elaborated by the different functional teams of the acquisition. These become more operational and larger than the acquisition teams but should at least partly consist of the same people to ensure consistency, sufficient management attention and learning. The planning follows a holistic integration planning process, whereas the discussion here focuses on the integration planning within the R&D and technology-related fields. As the technology integration is often not seen as a part of the integration process, state-of-the-art integration planning has so far not addressed this dimension. This gap is filled within the following section.

In order to appropriately address the technology integration in integration planning and to foster the short- and long-term innovativeness of the merging companies, the following aspects have to be addressed:

- Integration planning has to be adapted according to *acquisition type* and technology-based value-creation opportunities pursued;
- Integration planning has to be designed to foster the realization of the *technology-based value creation potential* and thus the achievement of future possibilities to realize planned and emergent technology-based value-creation opportunities;
- Integration planning has to consider the *contextual fit* between target and acquirer and thus the ability to successfully realize value creation.

The generic integration planning process (see Figure 6.48) consists of three phases: (1) the determination of the integration activities; (2) the assessment

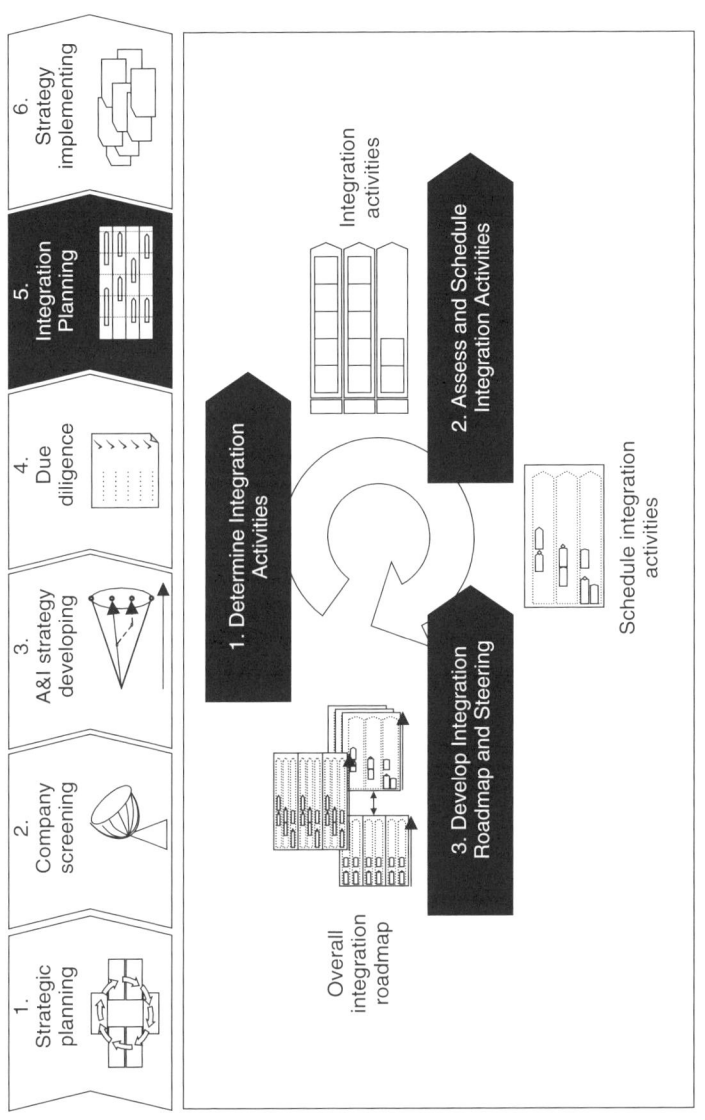

Figure 6.48 Integration planning process

and scheduling of the individual integration activities; and (3) the development of the integration roadmap and steering mechanisms.

6.3.5.1 Determine integration activities

In a first step each acquisition team has to determine which integration activities must take place in order to achieve value creation and to change the company configuration. Generally while three different layers of integration activities can be distinguished, two of these can be managed actively and one only passively (see Figure 6.49). The *task integration* is concerned with all value related activities within the functional layers. Thus within the R&D area activities to redefine value drivers, sustain existing value, capitalize on existing value and capture and create value have to be determined. Value definition is the development of a new technology and innovation strategy, whereas value sustaining is concerned with keeping the core competencies and pushing the daily business without focusing all resources on synergy realization. Value capitalizing and capturing is related to resource deployment and value creation is associated with innovation activities.

Besides these activities, the correct projects for changing the company configuration in order to support task-integration activities have to be determined. These projects concern organizational structure, change of supportive systems such as CAD, retention of key engineers, communication activities or general change-management activities. These activities are for now determined on a functional base but are finally managed in cross-divisional projects guided by the HR, IT, organizational, etc. teams.

The last activities which have to be considered but cannot be changed actively relate to the *socio-cultural* integration. Resource deployment and innovation activities require the acculturation of the underlying cultures and trust building. These developments have to be actively pushed by the supportive systems and task integration to enable a holistic change process in all three areas: task, supportive systems and socio-cultural integration.

6.3.5.2 Assess and schedule integration activities

In a further step, the determined integration activities have to be scheduled. The prioritization and planning of the different activities depends on their associated NPV, risk level and project type; thus these attributes need to be assessed first. Each integration project is assigned a specific objective to reach that is measured in financial figures or other attributes. These objectives are linked to the business value drivers. Furthermore, each project has a specific risk level derived from a revision of the risk assessment of the A&I strategy development and is categorized into a competitive-strategic or development-strategic project. Competitive-strategic projects contribute to the short-term competitiveness of the company, whereas development-strategic projects have a long-term impact, mostly concern various areas of the company and contribute to sustained and profitable growth.[19]

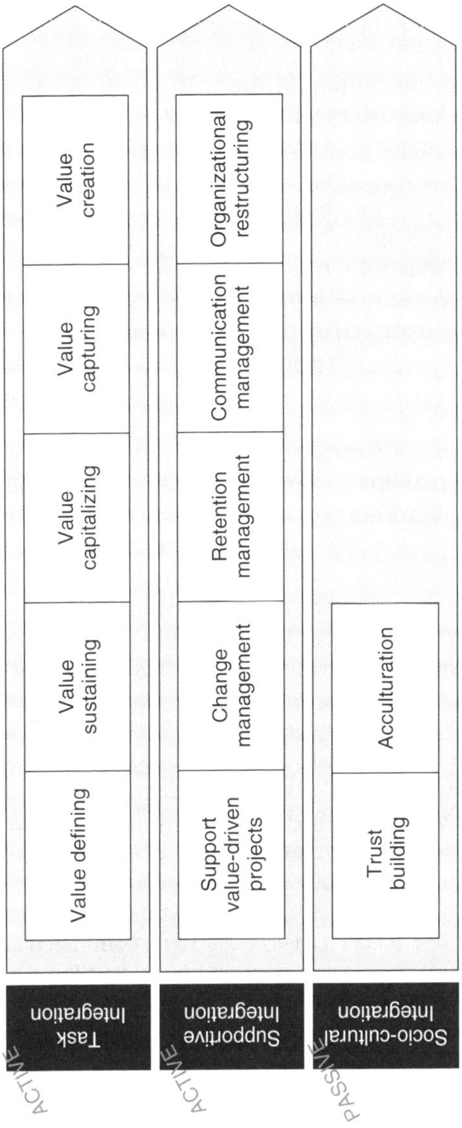

Task Integration (ACTIVE)	Supportive Integration (ACTIVE)	Socio-cultural Integration (PASSIVE)
Value creation	Organizational restructuring	
Value capturing	Communication management	
Value capitalizing	Retention management	
Value sustaining	Change management	Acculturation
Value defining	Support value-driven projects	Trust building

Figure 6.49 Three layers of integration activities

Subsequently, the different projects are prioritized within the different integration phases, according to specific focus and the resulting requirement for different projects of each. As already outlined the integration process can be separated into low-integration, restructuring and full integration phases, which can highly vary in their duration. Furthermore, a preparation phase preceeds the actual integration. Figure 6.50 shows an overview of which projects need to be assigned to which integration phase.

Once the initial plan for the task and supportive activities is scheduled, this plan has to be revised depending on whether there are sufficient resources to conduct all the activities as scheduled. Subsequently, each project has to be detailed. It has to be decided who will be responsible for carrying out the projects and how the team structure should look, for example whether there should be a task force team, or if the project will simply become part of daily business or is to be managed as a strategic project on the corporate level, which is often the case for development-strategic projects. Additionally, the staffing of the integration team has to be determined. It was reported as being favourable to have integration teams with employees from both companies, especially if the target is quite large. Furthermore, each project needs to be assigned an appropriate integration behaviour, which should also be manifested more broadly within an integration policy. Generally integration behaviour needs to be fast-paced and determined within the first two phases of the integration – during the low-integration and especially during the restructuring phase – whereas within the consolidation phase the behaviour should provide more slack for trust building and learning.

With respect to the technology integration this means that within the first two phases the focus is laid on retention of the key innovators and on achieving quick gains such as the leveraging of easily combined modules or cross-selling opportunities. Additionally there should be a focus on restructuring the underlying resource bases. Thus the strategic technology platforms need to be linked up, for example through co-location or the integration into networks or competence centres. Furthermore, boundary-spanning structures need to be established; new innovation processes need to be introduced; and the engineering and manufacturing infrastructure should be transferred and substituted. These large-scale changes need to be conducted and managed by a very experienced CTO or head of engineering, and led with rapid pace, a strong strategy and vision in mind and a clear leadership style supported by extensive communication. Thus within the restructuring phase the supportive activities take the main lead. Afterwards, within the consolidation phase, value capturing and creation are in the fore. The wheel of technology-based value creation starts spinning fostered by the realized company configuration. An example of an integration schedule within the product development area is provided in Figure 6.51. After detailing the schedule and also the interrelations between the different projects within each functional and supportive area, the overall integration roadmap has to be established.

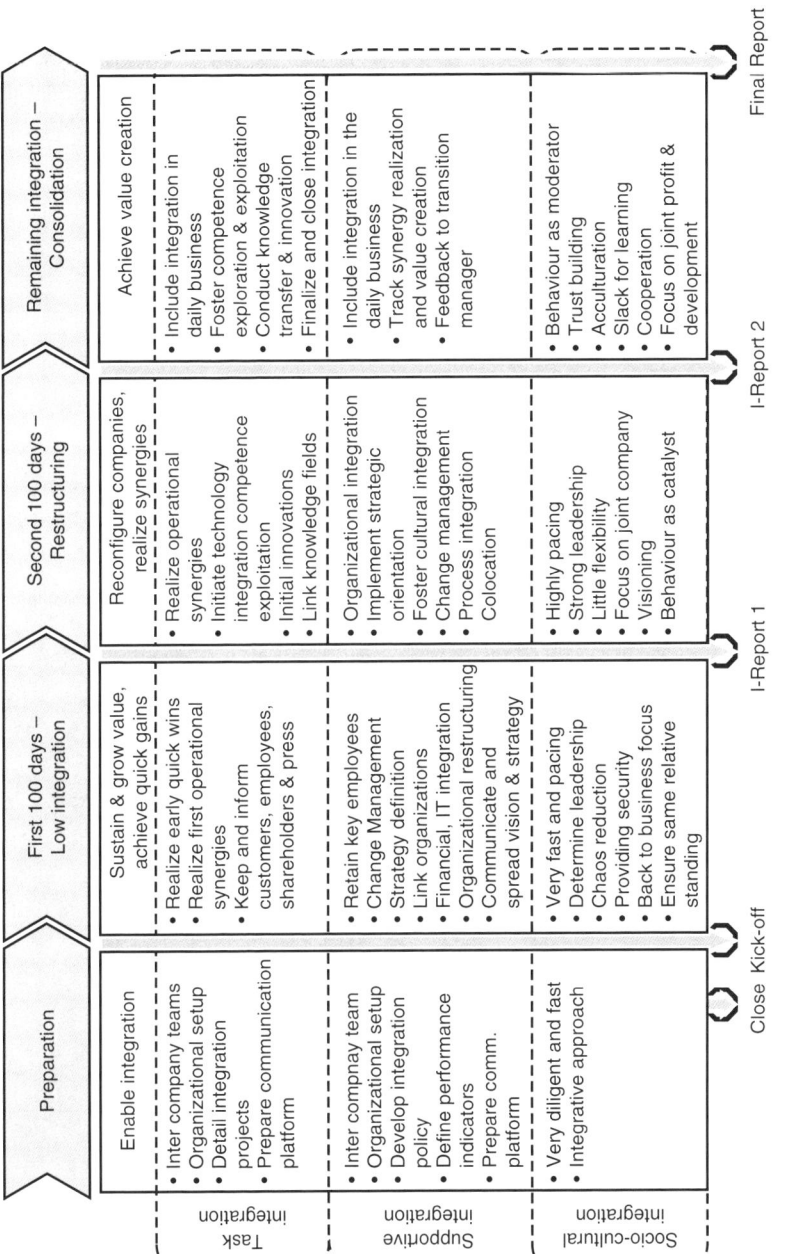

Figure 6.50 Scheduling activities within different integration phases

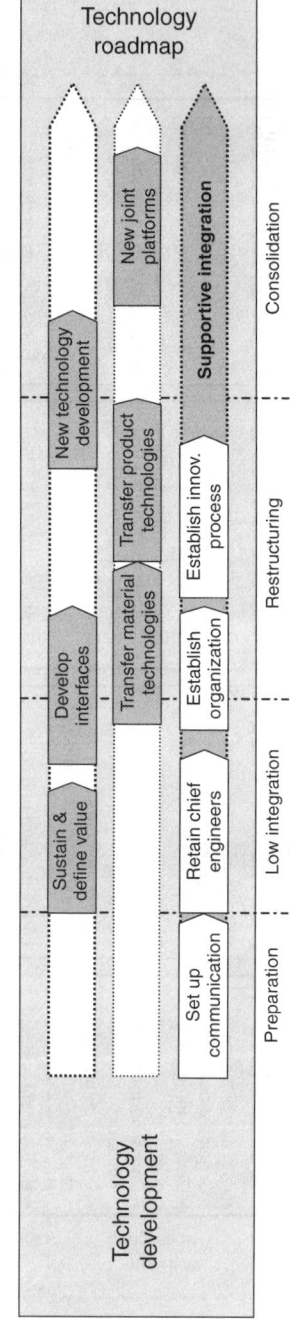

Figure 6.51 Integration schedule within product development

Strategic Acquisition and Integration Management

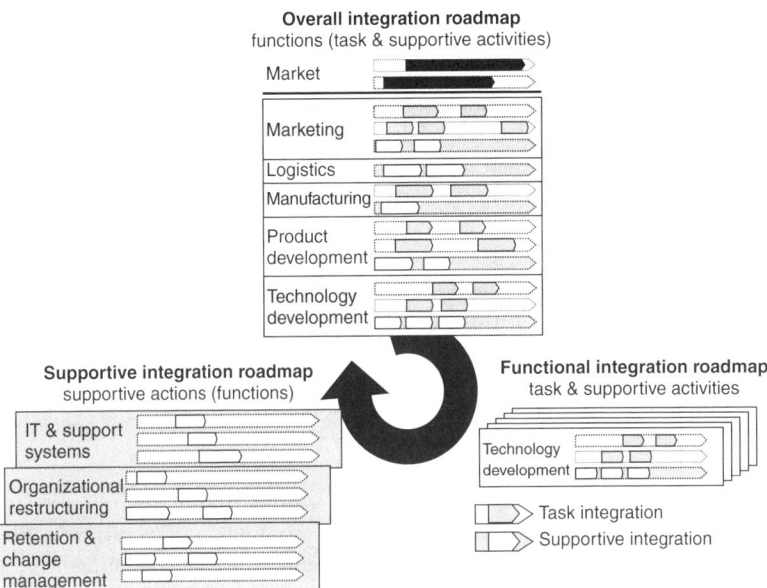

Figure 6.52 Three views of the integration roadmap

6.3.5.3 Develop integration roadmap and steering mechanisms

In order to ensure a holistic approach and guidance towards integration of the acquisition, in order to balance the workload and absorption of capacities and to avoid inconsistencies, an overall holistic integration roadmap including controlling and learning mechanisms should be developed as a management instrument. This has to comprise the perspective on the task and supportive integration and especially their interrelation. Thus the integration roadmap has three perspectives or layers. The first is the overall leadership view, which provides an overview of integration activities and their interrelations. The second view focuses on individual activities within one functional area as a guidance for each integration team and the third view shows only the supportive integration tasks within the whole company as a support for the boundary-spanning support teams, such as HR, IT, and the like. Figure 6.52 shows an example of such an integration roadmap and the different views and their interrelations.

Complementary to the integration roadmap the appropriate controlling and learning mechanisms have to be decided. The controlling tool should measure the integration success based on initially set objectives. It should be regularly applied and the bonus of the integration managers and also of the acquisition team should be influenced accordingly. This ensures that the acquisition team also sufficiently considers the integration aspects.

Another critical issue, especially when it comes to learning, is the active involvement of the acquisition team in the integration activities. This awareness is currently spreading in industry of a holistic acquisition and integration process which encourages companies to form acquisition and integration teams that are not only concerned with the financially dominated acquisition process but also provide guidance within the integration phase. This ensures that lessons learnt from the integration can be applied to selecting and evaluating future targets. Such permanent integration teams have been established at IBM and Cisco, for example.

Another structural approach which can be planned within or dependent from the integration planning is the establishment of permanent learning mechanisms. These can be an internal website on best practices and reports, a community of practice on M&A integration, or the like. One best practice example was established at Siemens (see Best Practice 13).

6.3.5.4 Structural and behavioural aspects

As mentioned above, integration planning is not a highly separate process at the end of the acquisition process. The applied considerations should take place throughout the process and should be actively managed by the transition manager, who is finally in charge of successfully realizing the integration.

From a behavioural perspective it is recommended that the target's employees should participate as early as possible in the integration planning process. Furthermore, integration planning should not be confined to a period of eight weeks, 100 days or even six months. Awareness has to be raised that the integration will continue for a long time after the acquisition and thus requires specific attention and efforts. Thus the technology integration, which is particularly difficult to achieve due to its dependence on supportive and socio-cultural integration, needs to be specifically addressed and managed as otherwise the wheel of technology-based value creation will never start spinning.

6.3.5.5 Conclusion

It can be concluded that a holistic integration planning of the integration activities within the different integration phases and the development of an integration roadmap and controlling and learning mechanisms have a significant impact on the long-term value creation in corporate acquisitions and thus enable the rise in innovativeness and the efficient deployment of capabilities with a reduced risk of destroying the corporate context and causing engineers to depart.

6.3.6 Strategy implementation

The final process of the acquisition and integration management is the integration process, which is more operational than strategic and thus not the

Strategic Acquisition and Integration Management 229

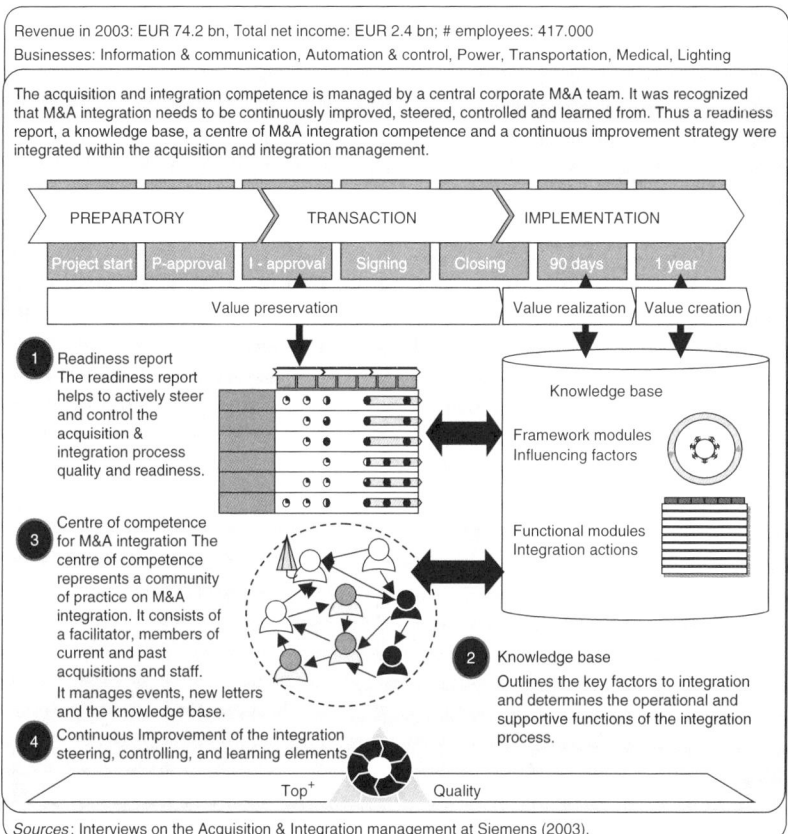

Best Practice 13   Integration steering, controlling and learning at Siemens

main focus of this book. It aims to implement the *formulated acquisition and integration strategy* and thus to achieve value creation and the strategic objectives. Therefore it is in effect the operational realization of the phased integration roadmap over subsequent years. The integration phase often has already started before the closing, with a preparation phase which overlaps the integration planning phase. This preparation phase, as mentioned before, is used to determine the individual projects, the teams and their objectives. The strategic level of the integration process aims to *steer and control the integration process* on the management, functional and supportive levels. This steering and controlling will support the acquirer to ensure the desired value creation, readapt the strategy with the still-decreasing information asymmetry and to learn from the acquisition. To foster the realization and emergence of technology-based value creation, the following aspects must be considered during integration:

- The integration process and behaviour have to be adapted according to the *acquisition type* and the pursued technology-based value-creation opportunities, following the defined integration approach;
- The integration has to merge the different corporate *contexts* and thus provide the ability to successfully realize the value-creation potential.

The operational integration process is concerned with task, supportive and socio-cultural integration within the different integration phases, following the planned actions within the integration roadmap. ABB has developed a best practice integration process which lasts from the pre-acquisition phase until long after the acquisition (see Best Practice 14).

The strategic level of the integration is mainly concerned with the steering and controlling of the integration progress (see Figure 6.53). This should be conducted with the previously developed and established controlling tools. An example for a controlling sheet of the integration progress is seen in

Best Practice 14 Integration process at ABB

Strategic Acquisition and Integration Management   231

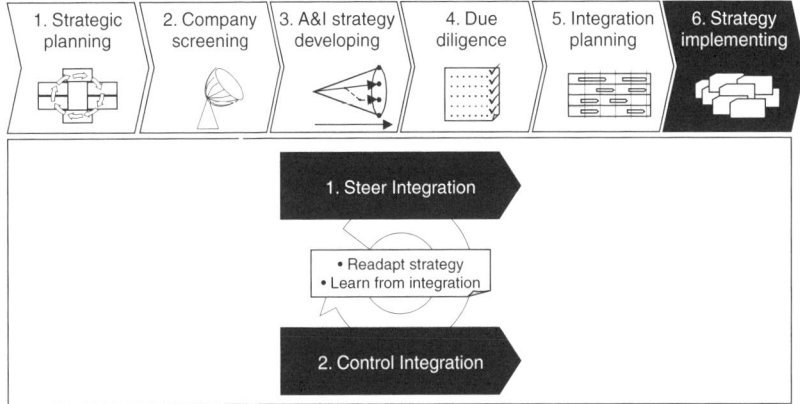

Figure 6.53   Strategy implementing

Figure 6.54. The measured integration performance over the subsequent years after the acquisition should influence the bonus not only of the transition manager but also of the people who decided upon carrying out the acquisition, such as the M&A team or the financial experts. In this way the integration issues and their long-term effects will be considered within ongoing and future acquisition processes. Furthermore, the integrated M&A person or consultant should gather the lessons learnt and establish a final integration report which needs to be provided as a feedback to the corporate M&A team and to future integration teams. This centralization of the management of the integration effort was observed in various companies. For example, HP, Cisco, GE and IBM all have separate integration teams which guide the process and provide support to the transition managers.

6.3.6.1   Structural and behavioural aspects

The operational integration conducted by the different integration teams is managed and guided by the transition manager for a period of about two years. He or she, the CEO of the business unit and the heads of the functional and supportive teams form the core integration team. Additionally, the integration team should be complemented by a member of the corporate M&A team or the internal consulting business in order to support the process with the appropriate project and integration management tools and to ensure appropriate feedback of the lessons learnt to the acquisition team and subsequent integration teams. Additionally, the integration team should be steered by a steering committee which ensures that integration remains top priority and is assigned full management attention. The CTO should also be part of the steering committee and eventually even the core integration team in order to ensure the appropriate consideration of technology integration and to promote boundary-spanning innovation projects. Figure 6.55 shows an example of an integration structure. The steering and controlling

	Variables	months	1 m	2 m	3 m	6 m	Correcting actions (1m)
Manufacturing	Performance (cost/unit; unit/time)	planned	−50	−10	−20	−150	
		actual	−40				
	Flexibility and capacity usage (utilization)	planned	90%	90%	90%	85%	
		actual	80%				
	Quality indications (FPY)	planned	+90%	+95%	+95%	+95%	
		actual	+85%				
	Employee satisfaction	planned			60%	80%	
		actual					
Engineering	Performance (time to market)	planned	−10	−10	−40	−50	
		actual	0				
	Retention rate	planned	90%	90%	90%	85%	
		actual	70%				
	Innovation rate	planned	+0%	+0%	+10%	+15%	
		actual	+0%				
	Employee satisfaction	planned			60%	80%	
		actual					

Figure 6.54 Integration performance measurement

233

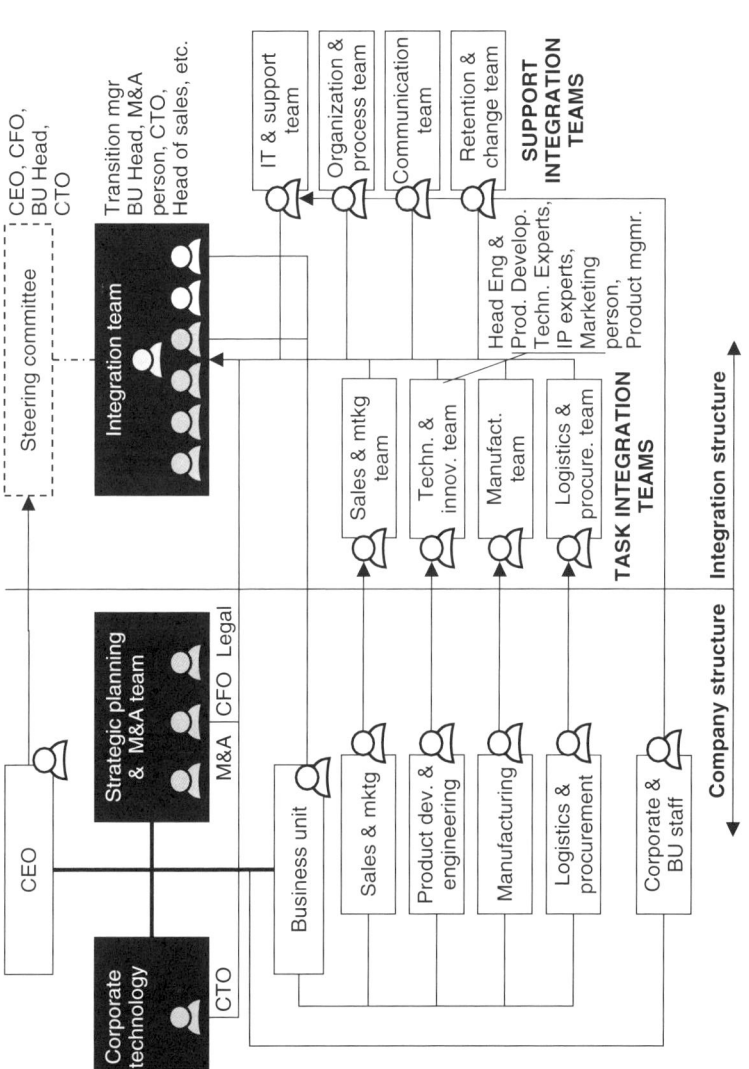

Figure 6.55 Example of an integration structure

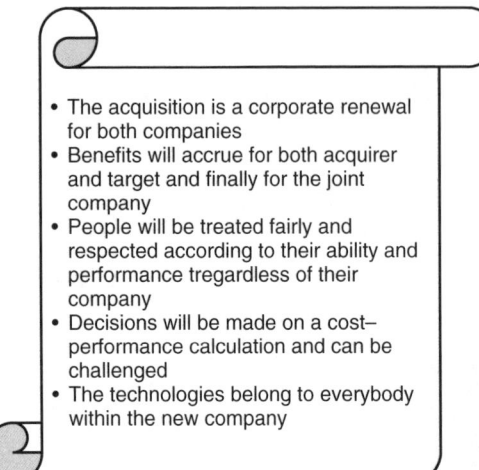

Figure 6.56 Integration policy

on the strategic level of integration should be done by the transition manager and reported to the integration steering committee. The behaviour of the integration teams should be according to the planned behaviour within the different integration phases. Furthermore, it should be supported by an integration policy (Figure 6.56).

6.3.6.2 Conclusion

With this approach to acquisition integration which is appropriately addressed by the transition manager, well planned in different phases and which considers technology integration as one integral element, the merged companies should be able to achieve increased innovativeness and efficiently redeploy the competencies. Additionally, through a more diligent steering and controlling of the integration progress lessons can be inferred and help to improve the subsequent acquisition performance.

6.4 Conclusion from Technology-based Strategic Acquisition and Integration Management

In the first chapter of this book it was outlined that companies struggle with becoming more innovative after an acquisition and with efficiently deploying the joint competence base. Thus, even though these technological synergies have become highly crucial for competitiveness in innovation-driven industries, they have not yet been mastered. Thus the two research questions guiding this book were posed: 'Which aspects in strategic acquisitions

1. Strategic planning	2. Company screening	3. A&I Strategy developing	4. Strategy validating	5. Integration planning	6. Strategy implementing
• Acquisition option should be considered as one integral and viable strategic path • Avoid opportunistic acquisitions • Avoid unrelated acquisitions • Technology and innovation management should be integrated in the strategy development • Conduct strategic planning with different time horizons • Ensure acquisition capability • Ensure willingness to change • Ensure established technology & innovation mgmt	• Define clear and competence-related need • Include technology intelligence in the data collection • Roughly evaluate technology based value creation potential • Roughly evaluate strategic fit including core competence fit • Roughly evaluate contextual fit • Include CTO & technology experts	• Analyse the company from a technological perspective • Actively identify technology based value creation opportunities based on merged resource bases • Assess value creation opportunities • Determine the integration plan based on the pursued technology-based value creation. • Pursue full integration unless only short-term value creation is pursued. • Determine company configuration to achieve technology based-value creation. • Include transition mgr	• Conduct a prospective technology due diligence • Validate potential for the occurrence of tbvcoby assessing the technology base • Validate potential for profitability of the tbvcoby assessing them • Validate potential for successful realization of the tbvcoby assessing the contextual fit • Holistically integrate the findings from the technology due diligence within the overall acquisition and integration strategy	• Determine activities within the task, supportive and socio-cultural integration to achieve value creation and a change in company configuration • Schedule technology integration activities according to the different integration phases: first focus on quick gains, establish boundary-spanning structures and then achieve long-term competence-based value creation • Integrate technology integration into an overall integration roadmap and steering system	• Actively steer and control the overall integration • Integration efforts should be managed by an transition manager • Learn from integration successes and failures • Conduct technology integration as planned • Focus on retaining the key innovators • Balance task, supportive and socio-cultural integration to achieve technology-based value creation

Figure 6.57 Summary of key issues necessary to achieve technology-based value creation

in innovation-driven industries determine the successful realization of technology-based value creation?' 'How can these aspects be successfully incorporated in the processes, structures and methods of the strategic acquisition and integration management in corporate acquisitions in innovation-driven industries?' Subsequently, theory was consulted to introduce the basics in technology and innovation management and in acquisition and integration management. Next the practice was investigated by case study research. Five acquisitions were analysed and an improved understanding of aspects impacting on technology-based value creation could be derived. This new understanding resolved the first research question. The second research question, which addresses applicable concepts for strategic acquisition and integration management to foster technology-based value creation, remained open. Thus, within this chapter, a technology-based strategic acquisition and integration management was developed. This mainly comprised an adapted strategic acquisition and integration process which was discussed on two levels. The first, rough level gave an overview of the acquisition and integration process and outlined the links between strategic, acquisition and integration, and technology and innovation management. Within the second, more detailed level the individual phases and layers of the acquisition and integration process were outlined. It was shown how these should be enhanced, changed or extended by adding new methodologies or changing the tasks or structures in order to ensure the appropriate consideration of technological aspects and to increase the likelihood of successfully realizing technology-based value creation. The main issues which have to be considered within the respective phases to ensure that the joint companies can increase their innovativeness and that they will be able to efficiently redeploy their competencies are summarized in Figure 6.57.

The proposed technology-based strategic acquisition and integration management model cannot assure the increase in innovativeness or the successful internalization and deployment of the technology bases; nonetheless it is definitively supportive of achieving these goals and can serve as a useful M&A decision-making guideline. Thus this concept resolves the lack of concepts for successfully mastering technology-based value creation in corporate acquisitions in theory and practice.

7
Management Principles

Before closing this book, the practitioners are addressed one more time. Within this last chapter the main findings from this book will be summarized and formulated for managers and decision-makers as ten management principles. The different principles should represent not empirically profound findings but rather critical guidelines to be addressed by management. Thus the elaboration of the principles is not thoroughly argued and it is written in management-style language. The management principles are divided into understanding of technology-based value creation in corporate acquisitions, applicable concepts for mastering it and implementation of these concepts.

7.1 Management Principles for a New Understanding

The management principles for a new understanding summarize the critical points which managers need to understand if they wish to increase their ability to achieve technology-based value creation after acquisitions. These four fundamental and general statements are partly interrelated and partly contradict the common understanding in theory and practice.

7.1.1 Management Principle 1: Wheel of technology-based value creation

It is important to understand that the initial cross-selling or easily to achievable innovation successes after an acquisition do not guarantee further success; actually it is the other way round. Whereas initial successes can be best achieved in keeping the target quite separate, subsequent more long-term-oriented technology-based value creation requires a very different,

> Technology-based value creation in corporate acquisitions occurs in two subsequent phases. In an initial phase only easy-to-achieve quick gain innovation opportunities occur whereas in the next phase an increased and long-term innovativeness and efficient resource redeployment iterate, emerge and promote each other. Managers have to understand this cycle and manage it appropriately.

more tightly linked integration approach and thus an innovation- and cooperation-fostering company configuration and the appropriate integration behaviour. Thus initial quick gains and long-term emergent technology-based value creation have to be pursued differently. This implies that managers should not be satisfied with quick gains and the retention of a distant company configuration but should actively guide the integration process in phases and adapt the company context according to the desired value creation. Furthermore, managers have to become aware of the emergence particularly of long-term technology-based value-creation opportunities and thus, instead of directly planning and implementing innovation opportunities, they should focus on implementing a context which enables the occurrence of efficient resource deployment and valuable innovation opportunities.

7.1.2 Management Principle 2: Innovativeness cannot be bought

> Innovativeness cannot be bought by acquiring a highly innovative company. If the acquirer is not innovative itself, the target will lose its own innovativeness rather than foster the joint companies' innovation capability.

Many acquirers today intend to become more innovative by acquiring a highly innovative company. However, as the long-term innovativeness of the joint companies is dependent on their merged context, the acquirer as well as the target has from the beginning to provide innovation-fostering contexts which can be well merged and are thus quite similar. If one of the companies does not have an innovation-promoting context, the merged context cannot ensure the innovativeness and ability to efficiently redeploy resources for the joint companies. This implies that an acquirer first has to be innovative itself before fostering own innovativeness by acquiring an innovative target. Furthermore, it is not the best-in-class target but the one which fits best with the acquirer that has to be sought and internalized.

7.1.3 Management Principle 3: Willingness to change as key initial condition

> Without the acquirer's commitment and willingness to change, the formation of a new joint company based on merged resource bases and the achievement of short- and long-term technology-based value creation is not possible.

Short- and long-term technology-based value creation can be achieved only if the merged company is based on one joint resource base and accordingly has one merged corporate context. This new corporate context is different from that of the acquirer or the target as it comprises a whole new, enhanced and enlarged resource base. Thus not only does the target need to adapt to the new joint context but also the acquirer. Therefore, in order to successfully realize technology-based value creation the acquirer has to be willing to change. This willingness has the positive side-affect that the acquisition will be assigned sufficient resources and paid much management attention, and that it represents little difference in relative standing between the acquirer and the target. This implies that an acquisition has to be viewed as an opportunity to change and needs to be accompanied by change management, a high sense of urgency, communication and the like in order to succeed.

7.1.4 Management Principle 4: Creative destruction of innovation context

> The successful realization of short- and long-term technology-based value creation requires the target's full integration and thus the destruction of its existing organizational, cultural and dominant logic context risking loss in innovativeness and the departure of key innovators. These risks have to be compensated by quickly providing the new joint innovation-fostering context.

In particular long-term technology-based value creation requires tight cooperation and communication as well as trust building. This again can be achieved only within one joint innovation and cooperation fostering corporate context. This tight integration, however, destroys the existing and often innovation-fostering context of the target company. This is associated with the risk of the departure of key innovators and the decrease in innovation rate. Instead of avoiding this risk by keeping the target separate, as proposed by several authors and practitioners, the risk has to be faced and averted by quickly providing the new joint context which builds the basis for new joint innovativeness and by implementing this in a dynamically paced and determined way.

7.2 Management Principles for a New Concept

The management principles for a new concept summarize the key findings practitioners need to consider when designing, directing and developing management concepts for strategic acquisition and integration management which aim to foster the achievement of technology-based value creation.

7.2.1 Management Principle 5: Technology concerns as basic fundamental

> Technology is not only a concern in innovation-driven acquisitions but a key factor in all strategic acquisitions. Thus the technology-based strategic acquisition and integration management has to become the foundation of all acquisitions in innovation-driven industries.

An increase in innovativeness and the efficient deployment of capabilities is not only a concern in innovation-driven acquisitions but a key strategic issue in all strategic undertakings in innovation-driven industries. Thus the long-term success rate of strategic acquisitions with the main objective of acquiring a market share or reaping scale effects is in the end also determined by the joint companies' innovative products which are then sold and efficiently manufactured. Thus technology- and innovation-related concerns need to be addressed within all strategic acquisitions. This is particularly emphasized when in innovation-driven acquisitions technology concerns are regarded as important but lack mastery, whereas in other acquisitions technology concerns are neither regarded nor mastered. This implies that the acquisition and integration management for all strategic acquisitions in innovation-driven industries has to consider technology aspects.

7.2.2 Management Principle 6: Holistic, systemic and integrative concept

> The technology-based strategic acquisition and integration management has to be designed and developed in a holistic, systemic and integrative way in order to ensure the appropriate consideration of the technology- and innovation-related concerns.

The specific elements of the technology-based strategic acquisition and integration management such as the technology due diligence, the identification of technology-based value-creation opportunities or the planning of technology integration should not be applied as stand-alone concepts, but need to be one integrated element of the general strategic acquisition and integration management. Furthermore, the ideas and concepts have to be holistic and thus need to be supported by appropriate processes, structures, methodologies and behavioural implications. Additionally, it was observed that the modular character of the concept consisting of different elements which can be applied throughout the process is highly accepted and easily applicable. Last but not least, it became apparent that the best technology due diligence is useless if the results are not integrated within overall strategy-making. Thus the elements of technology-based strategic acquisition and integration management need to be tightly integrated and linked and add up to one holistic acquisition and integration management for innovation-driven industries.

7.2.3 Management Principle 7: Prospective perspective

> The specific elements of the technology-based strategic acquisition and integration management need to pursue a prospective perspective on the role of technology and innovation in corporate acquisitions.

State-of-the-art concepts that consider technology-related aspects within acquisition and integration management focus on the past and current value of technologies and their context. This approach neglects the fact that the value of a technology is determined by its future rent generation or associated future cash flow streams. Thus the new concepts to deal with technology-related issues in acquisition and integration management have to identify, assess and plan technologies and competencies according to their future use and associated value creation potential.

7.3 Management Principles for Implementation

The management principles for implementation summarize the key findings practitioners need to consider when implementing and directing management concepts for strategic acquisition and integration management.

7.3.1 Management Principle 8: Implementation requires a new awareness of acquisitions

> Successful implementation of technology-based strategic acquisition and integration management requires a new and shared awareness of the role of acquisitions as one viable strategic path for corporate renewal in strategic management.

The implementation of technology-based strategic acquisition and integration management requires the acquirer's shared awareness about acquisitions. Thus acquisitions should be seen as one viable strategic path for achieving corporate renewal. They should not be the extension of the existing company but provide the opportunity to form a completely new joint company with a newly configured resource base and a new context which offer the opportunity to create new values.

7.3.2 Management Principle 9: Implementation requires a new awareness of technology and innovation

> Successful implementation of technology-based strategic acquisition and integration management requires a new and shared awareness of the role of technology and innovation management as an integral and dominant element in strategic and in acquisition and integration management.

The implementation of technology-based strategic acquisition and integration management can be fostered by a shared awareness of the technology and innovation management within a company. It has to be understood that technology and innovation management with all its concepts, processes, methods and structures fundamentally contributes to the mastery of innovativeness and the ability to efficiently redeploy competencies and thus achieve value creation and competitiveness. Thus technology and innovation management should be an integral aspect of all strategic decisions within a company and thus also within corporate acquisitions. But not only does the strategic and acquisition management need to embrace the technology and innovation-related aspects but the reverse is also important. The technology and innovation management so far hardly considers the acquisition option as one viable method for internalizing and redeploying competencies. This perspective needs to be changed; the CTO should see acquisitions as an opportunity rather than a threat and should use these options as a technology- and innovation-strategic path and seek a stake in and responsibility for implementing the decision.

7.3.3 Management Principle 10: Implementation requires critical self reflection

> Successful implementation of technology-based strategic acquisition and integration management requires a critical self-reflection on past acquisition performance particularly measured in innovativeness and efficient resource deployment.

This management principle encourages managers to critically reflect on their past acquisition and integration performance in particular regarding the innovativeness and the achievement of an efficient redeployment of the internal resource base. This investigation will raise the awareness of the critical impact of acquisitions on the company's innovation performance and persuade management of the usefulness of implementing technology-based strategic acquisition and integration management.

Appendix A – Value Creation in Corporate Acquisitions

This appendix will provide an overview on the state of the art regarding value creation in corporate acquisitions. It aims to answer the question: which aspects and criteria are relevant for achieving value creation within corporate acquisitions? This topic is important for understanding the general success factors in corporate acquisitions. Furthermore, these findings will be used to derive a framework for the case study analysis.

A wide variety of researchers from various different schools have tried to identify the main relevant criteria that bring about successful acquisitions. They all apply different methods for analysing the success of corporate acquisitions, investigate different phenomena, find different explanations and infer different implications for practice.[1] Unfortunately, however, very few researchers have widened the scope of their research by investigating the interrelation of the different aspects observed. Exceptions are the representatives of the process school who try to cover some of the aspects derived from other schools. Furthermore, Larsson and Finkelstein (1999) have conducted an extraordinary attempt to merge the various schools and provide a more holistic understanding of the subject of corporate acquisitions. They were rewarded for their approach by Kathleen Eisenhardt, stating that: 'This paper is exciting because it synthesizes several theoretical perspectives into an integrative model and addresses a very significant topic – mergers and acquisitions – with a sharp eye towards clear managerial relevance and with innovative methods. I expect it to become a defining paper in M&A research' (Larsson & Finkelstein, 1999: 1).

Haspeslagh and Jemison (1991) have proposed a well-accepted categorization of schools investigating the sources of value creation in corporate acquisitions, which will be used here as well:

- Capital markets school
- Strategic school
- Organizational behaviour school
- Process school.

Within this literature review an attempt is made to provide the reader with an understanding of aspects relevant for value creation and to identify criteria which are particularly relevant to technology-based value creation.

A.1 Capital Markets School

The capital markets school is mainly concerned with the effect of acquisitions on net wealth gains of the shareholder. Financial economists, applying the 'market for corporate control'[2] perspective, understand acquisitions as a contest between competing management teams for the control of corporate resources (Jensen & Ruback, 1983: 6).[3] They apply the event study methodology[4] to analyse the changes in share price for both the acquirer and target in a short time period around the acquisition event. They base their argumentation on the efficient market hypobook,[5] saying that the expected future cash flows associated with the growth strategy are immediately incorporated into the share price of the acquiring and target firms by the rational and efficient players of the capital market.[6]

This, however, implies that the financial economist generally assumes that the market is efficient, which means that shareholders can understand the way in which a firm's strategy will evolve and which cash streams will be associated with it. Furthermore, they expect the management to act on behalf of the shareholder rather than in their own interest – the typical agency problem.[7]

Based on these assumptions researchers have found various results. Whereas some argue that additional wealth is created by acquisition and thus social value is generated,[8] others have not found positive results regarding gains for the shareholders.[9] Generally however, it was found that the *acquiring firm neither gains nor loses* (0–4 per cent) whereas the acquired firm often earns even significant gains of up to 20 to 30 per cent.[10] For more detailed overviews on gains from a capital market perspective see Jensen and Ruback (1983), Sirower (1997) or Rockholtz (1999a) and Carper (1990: 809). The impact factors on wealth creation vary and include the *bidder's approach* (tender offer vs. merger), the *financing mode* (cash vs. stock financing) or *numbers of bidders*.[11] Furthermore, the impact of acquisitions on the systematic and unsystematic risk by using the Capital Asset Pricing Model (CAPM)[12] were investigated and showed that generally the unsystematic and thus corporate specific risk increases, whereas systematic risk can be reduced in related acquisitions (Lubatkin & O'Neill, 1987).

Researchers have identified various reasons for potential value creation but also for the observed losses. According to Shleifer and Summers (1988) gains are derived from a new relation between the shareholders and the management. By exchanging poor management which is not acting in the interest of the shareholder *agency costs* can be reduced. Another explanation for potential gains has roots in potential *tax manipulation*; however, it was shown that these considerations, even though they can impact on the acquisition decisions,[13] are not relevant to management.[14]

The explanations provided for the lack of value creation are also partly rooted in the agency theory. Companies often have remaining free cash flow[15] which, according to financial economists, has to be paid to the shareholders, which then have the opportunity to invest elsewhere. However, instead this free cash flow is used to conduct an acquisition, motivated by the personal interests of the management to gain power and increase their salaries (Haspeslagh & Jemison, 1991: 294). Thus the wealth creation associated with acquisitions is absorbed by management rather than shareholders. Another line of argument says that wealth creation is not achieved due to the *hubris* of managers[16] or their *mistakes in assessing the potential for future value creation*.[17] Balakrishnan (1988) argues differently; he states that no value is created as the investors have already anticipated the acquisition as the logical step to follow the pursued strategy.

Even though the capital market school is the dominant stream of research especially in the US,[18] its underlying assumptions and the applied methodology are subject to criticism. As described, the findings are based on the claim that markets are efficient, investors rational and strategies and their related cash flows can be anticipated. This 'contradicts a good deal of what we now about strategy, how it is developed, and how it evolves' (Haspeslagh & Jemison, 1991: 298). As is well known from Mintzberg and Lampel, (1999), Eisenhardt (1999) and many others, strategy is not deterministic but rather an evolving process. Furthermore, the applied methodology analysing the changes in the share prices 'provide a convenient but hazardous and single-minded measuring stick ... which is tied to short-term performance' (Haspeslagh & Jemison, 1991: 300).

Thus the research on corporate acquisitions from a capital market perspective does not address long-term value creation, such as the improvement of competitiveness, due to an increase in the innovation rate or an efficient deployment of capabilities.

A.2 Strategic Schools

The strategic schools, as the dominating ones in corporate acquisitions in general, have multiple roles. The research in strategy content and process builds the foundation for strategic acquisition and integration management. Furthermore, the strategic schools help explain the different acquisition motives. Last but not least, researchers from the strategic schools have identified aspects relevant for successful value creation in corporate acquisition, which will be elaborated within this chapter.

Whereas the capital market school investigates the effect of acquisition on value creation in a very general way, the strategic school focuses on the *impact on the individual company*. The researchers[19] analyse the impacts of *target size, industry-relatedness, acquisition experience, profitability*, etc. on the associated value creation. Several[20] apply the event studies to measure the success of the acquisition, as do the representatives of the capital market school.

The majority of the researchers focus their attention on the *relatedness of the acquirer and the target*. This research stream originated with the research on diversification which was highly influenced by the concepts of Rumelt (1974), who found that related-constrained firms were more successful then related-linked firms, which outperformed unrelated diversified firms. The following stream of literature provides very mixed conclusions. Several researchers argue in favour of related acquisitions,[21] others oppose, arguing that unrelated acquisitions have the potential to improve income stability, reduce the likelihood of bankruptcy and increase the market value of debt of the combined firms.[22]

Several researchers[23] provide evidence that *related mergers generally gain more value* than unrelated ones. Furthermore, they[24] describe the sources of value creation such as economies of scale and scope or market power economies. Another important impact derives from similarities in business style,[25] which is referred to as a '*dominant general management logic*' (Prahalad & Bettis, 1986). Others could not find any support for improved gains in related acquisitions[26] or have even observed a negative effect of the relatedness.[27]

Thus the results are inconclusive at best. Barney (1988) has added another aspect which determines the value creation associated with acquisitions. He argues that 'only when bidding firms enjoy *private and uniquely valuable synergistic* cash flow with targets, *inimitable* and uniquely valuable synergistic cash flow with targets, or *unexpected synergistic* cash flows, will acquiring a related firm result in abnormal returns for the shareholders of bidding firms' (Barney, 1988: 71).

Another line of research within the strategic school has focused on the impact of *acquisition experience* on acquisition performance.[28] Haleblian (1999: 29) has found a U-shaped relationship between organization acquisition experience and acquisition performance. Whereas learning occurs during the first acquisitions of a company, more experienced acquirers discriminate between their acquisitions – the organization experience effect. Rovit (2003) has also found that acquisition experience is conducive to acquisition performance, especially when feedback systems and opportunities for learning were installed. Thus it can be concluded that in general acquisition experience using feedback systems and not getting trapped in routines are supportive to superior acquisition performance.

A.3 Organizational and Behavioural School

The behavioural and organizational school, which has attracted many researchers especially in recent years, addresses two main topics. On the one hand, the impact of acquisitions on the individual especially from the target company is analysed and concepts for mastering this human side of acquisitions are proposed. On the other hand, the influence of corporate cultures and their fusion on acquisition success is analysed.

The analysis of the acquisition's impact on the individual person comprises the impact of the transaction on the behaviour of personnel and their reaction as part of a crisis situation. Several authors[29] have analysed employees' typical feelings of conflict, tension, alienation, career uncertainty, behavioural problems, stress, loss of productivity, concerns about financial security, geographic relocation, and co-worker trust. Furthermore, the collective reaction of personnel to the acquisition as a situation of crisis was assessed. The authors describe a process beginning with shock, defensive retreat, acknowledgement and finally a move to adaptation (Marks, 1982; Devine, 1984).

Other authors[30] have focused on the impact of acquisitions on *job position and turnover rate*. These representatives from human resource literature view a high turnover rate as a negative impact of acquisitions in terms of loss of knowledge and a symptom of a decaying work environment. They have found a high rate in management turnover mainly within the acquired firm. Walsh and Ellwood (1991) have shown that this phenomenon is not related to poor prior performance of the target company, indicating that the turnover is not due to the pruning of underperforming management (Weber & Camerer, 2003).

In order to master these issues of decaying working environment, demotivated personnel and high turnover rate, concepts for a *fair and conflict-resolving integration approach* have been developed. Intensive and honest communication is seen as one major criterion supporting socio-cultural integration.[31] Furthermore, the introduction of a shared symbolism is a powerful tool for fostering integration.[32] Subsequently, a cultural or human resource due diligence is introduced as a complementary methodology supporting the acquisition process (Greengard, 1999).

Besides the impact of acquisitions on the individual, the role of the corporate culture is investigated. Several authors[33] have described the impact and relevance of *cultural and organizational fit*[34] and claim their diligent consideration within the acquisition and integration process. It was shown that a certain level of relatedness in cultural and organizational fit are conducive to integration success. Furthermore, it was argued that the level of change on the group or business unit level versus the corporate level has an impact on integration (Napier, Simmons, & Stratton, 1989). The process of integrating different organizational cultures has also been analysed (Nahavandi & Malekzadeh, 1988) and an appropriate design (Birkinshaw & Hakanson, 2000) and integration behaviour (Chaudhuri & Tabrizi, 1999; Steingraber, 2001) for the acquisition and integration processes are discussed.

It can be concluded that acquisition success is highly dependent upon two important factors. On the one hand the integration approach and behaviour has to be tailored to reduce the negative impact of acquisitions on the individual person and to allow smooth cultural integration. On the other hand acquisition decisions need to diligently consider the impact of different organizational cultures on integration success. Unfortunately, however, up to now very little is known on the interrelation of the organizational and behavioural as well as other influential aspects in acquisition management, such as the economic benefit or process structure. Furthermore, the impacts of different acquisition types and motives and also technology-based value creation are not considered.

A.4 Process School

The process school, dominated by Haspeslagh and Jemison (1991) and Jemison and Sitkin (1986a; 1986b) and also more recent researchers such as K. Lucks (2002), takes a more integrative perspective, comprising aspects of the economic, behavioural and organizational and strategic schools. These researchers are concerned with the impact of the design of the acquisition and integration management and its processes on the economic performance of acquisitions. The acquisition and integration process and associated decision-making becomes the focus. Thus several authors[35] have developed process frameworks for the acquisition and integration process. Other researchers[36] have identified the main problems and barriers during the acquisition and integration process hindering successful value creation. The most known article on the topic was written by Jemison and Sitkin (1986a) who describe the impact of activity segmentation, an escalating momentum, exceptional ambiguity and management system misapplication.

The research focus and content of the various researchers is highly diverse and partly unrelated. Some authors focus only on the impact of one specific phase of the acquisition and integration process, such as the due diligence (Berens, Hoojan, & Strauch, 1999) or the integration process (Finkelstein, 1986; Steingraber, 2001). Others focus on the individual decisions to be taken during the acquisition and integration process, for example several authors discuss the decision on the integration approach and its impact (Shanley, 1987; Pablo, 1994).

It can be concluded that the literature on the process perspective of acquisition and integration management is convincing due to its holistic approach; however results are still highly diverse and unproven. As Sirower (1997: 156) points out: 'this literature is still in the theory-building stage, however, with little empirical evidence'.

A.5 Model of Understanding: Framework of the Main Aspects Affecting Value Creation in Corporate Acquisitions

Reflecting on the literature review above of the four main schools which are concerned with the aspects relevant for value creation in corporate acquisitions, a certain pattern of aspects can be identified. It can be observed that the researchers have found that different types of value creation are dependent on three main aspects (see Figure A.1):

- Acquisition type
- Initial conditions
- Strategy processes.

The first main impact factor is the *acquisition type*. The representatives of the capital market school and the strategic school have pointed out that not all acquisition types lead to the same amount and way of value creation. The acquisition type is described by variables which characterize the transaction between the buying and the acquired company. These variables are, for example, the relative size of the target to the acquirer, the business and industry relatedness between the two companies and aspects such as financing mode, bidder's approach, and so on.

School	Authors	Acquisition Type	Initial Conditions	Strategy Processes		Value Creation
				Acquisition process	Integration process	
Capital Markets School	Jensen & Ruback 83	■				■
	Jarrel, Bricklea, Netter 88	■				■
	Banerjee, Owers 92	■				■
	Sirower 94	■				■
	Healy, Krishna et al. 979	■				■
	Lubatkin & O'Neill	■				■
	Datta, Pinches et al. 92	■				■
	Schleifer, Summer 88					■
	Roll 86				■	■
	Lubatkin 83				■	■
Strategic School	Kitching 69	■				■
	Kusewitt 85	■				■
	Fowler, Schmidt 89	■				■
	Duncan & Savill 97	■				■
	Lubatkin, Calori, et al. 98	■				■
	Salter, Weinhold 79	■				■
	Lewellen 71	■				■
	Higgins, Schall 75	■				■
	Leontiades 86	■				■
	Bettis 81	■				■
	Christensen & Montgomery 81	■				■
	Bettis & Hall 82	■				■
	Singh & Montgomery 87	■				■
	Shelton 88	■				■
	Ramaswamy 97	■				■
	Seth 90	■	■			■
	Davis 68		■			■
	Prahald & Bettis 86		■			■
	Haleblian 99		■			■
	Gagnon, Sheu 00		■			■
	Finkelstein & Haleblian 02		■			■
	Rovit 03		■			■
	Hayes 97					■
	Sutton 83					■
	Marks & Mirvis 85					■
	Sales, Mirvis 85, 98					■
	Buono, Bowditch, Lewis 88					■
	Marks 82					■
	Devine 84					■

Figure A.1 Literature overview of aspects relevant to value creation in corporate acquisitions

School	Reference	Col1	Col2	Col3	Col4	Col5
Organizational & Behavioural School	Walsh 88, 98					■
	Hambrick, Cannella 93					■
	Kruch, Hegarty 97					■
	Davis, Nair 03					■
	Walsh, Ellwood 91					■
	Sinetar 81				■	■
	Bastien 87				■	■
	Napier, Simmons et al. 89				■	■
	Schweiger, DeNisi 91				■	■
	Dooly, Zimmermann 02				■	■
	Hirsch 87				■	■
	Schneider, Dunbar 87				■	■
	Greengard 99			■	■	■
	Buno & Lewis 89		■			■
	Datta 91		■			■
	Mendonca, Kanungo 94		■			■
	Testa, Mueller, Thomas 03		■			■
	Franck 90		■			■
	Cartwright, Cooper 93		■			■
	Napier, Simmons et al. 89				■	■
	Nahavandi, Malekzadeh 88				■	■
	Birkinshaw, Hakanson 00				■	■
	Chaudhuri, Tabrizi 99				■	■
	Steingraber 01				■	■
Process and other M&A related School	Haspeslagh & Jemison	■	■	■	■	■
	Jemison, Sitkin 86 a,b			■	■	■
	Lucks 02			■	■	■
	Bastien, Ven 86			■	■	■
	Haspeslagh, Farquhar 87			■	■	■
	Napier 89			■	■	■
	Duhaime, Schweink 85			■	■	■
	Shrivastava 86				■	■
	Haunschild, Davis-Blake et al.94			■	■	■
	Berens, Hoojan, Strauch 99		■			■
	Finkelstein 86				■	■
	Shanley 87				■	■
	Pablo 94				■	■

Figure A.1 Continued

The second main impact factor is the *initial conditions* which characterize the two companies and the attribute and relation of their characteristics. Authors from the strategic and organizational and behavioural schools have identified attributes of the buying and acquiring companies which are conducive to value creation. Additionally, they have found that a certain relationship between some attributes is especially supportive of value creation. Thus acquisition experience of the acquirer and a cultural and organizational fit between the two companies support value creation. Furthermore, similarities in the dominant general business logic are also conducive to value creation between the two companies. Another criteria mentioned is the potential for unique, difficult to imitate and valuable synergies. It is argued that only these types of hidden synergies are the real gains from corporate acquisitions.

The third main impact factor is related to the *strategy processes* such as the acquisition and integration processes. The importance of the design and tasks of the strategy processes is mainly derived from the organizational and behavioural and process schools. It can be concluded that the process design and content, the decisions made, people involved and methods applied have a significant impact on the overall acquisition success.

The three criteria mentioned above will form the framework to analyse the case studies and to identify the criteria relevant for technology-based value creation in corporate acquisitions.

Appendix B – Checklist: Technology Due Diligence

The technology due diligence as described within this book is the prospective validation of the successful realization of the future technology-based value-creation potential of the joint companies. Thus a mere retrospective analysis of the technology base or the product and technology pipeline is not sufficient to build a picture of the future. Thus the proposed checklists extend the validation phase and cover aspects of analysis and planning. Thus in a first step the checklist for analysing the company is outlined. Secondly, the main aspects to consider during the development of the joint innovation and technology strategy and thus the development of the joint technology and innovation roadmap are listed. Thirdly, the validation checks which address the potential of occurrence of technology-based value-creation opportunities, the potential of their profitability and the potential for joint successful realization are described. Thus this checklist is not only confined to being used during the technology due diligence but is to be used throughout the whole acquisition and integration process.

B.1 Screen Company

Company screening evaluates whether the potential target matches the acquirer and its strategy. Thus the following investigations on the company level can be conducted.

Evaluate value-creation potential

The potential value-creation opportunities:

- Are unique and valuable to the combination of the two companies;
- Are based on the combination of strong resource bases;
- Are in line with the long-term strategic direction;
- Can be achieved with reasonable efforts.

Evaluate strategic fit

The potential target and the associated value creation:

- Contribute to short- and long-term objectives;
- Match the pursued strategic goal and path;
- Adhere to the corporate and business unit strategy;
- Adhere to the different functional strategies such as marketing, HR, financial and also technology and innovation strategy.

The strategic fit evaluation also comprises a specific technology strategic fit evaluation. It consists of two elements: the fit of the target's technological attractiveness; and its match.

Evaluate technology strategic fit

The attractiveness of the core technologies of the target is measured by their:

- Maturity;
- Functional Scope;

- Scalability;
- Robustness.

The target's core technologies match to the acquirer if they:

- Have the potential to be a core competence;
- Support core competencies;
- Fit with technology portfolio.

Evaluate contextual fit

- Attractive and matching corporate cultures;
- Attractive and matching organizational structures;
- Attractive and matching dominant business logic.

B.2 Analysis of the Company

The following questions have to be posed when analysing a potential target:

- What are the target's resources and product and process technologies?
- What are the target's core competencies?
- What are the target's strategic technology platforms, functions and underlying knowledge?
- What are the target's business fields, markets and customer needs?
- How are the resources integrated in the current products and linked to each other?
- How are the technologies deployed within the value chain?
- How are resources planned to evolve and change over time including an investigation of what the target has in its product and technology pipeline?
- Where are the main strengths and weaknesses?
- What are the main trends influencing the resource base?

B.3 Identification of Technology-based Value-creation Opportunities

In order to validate the value creation potential, the future technology and product roadmap or pipeline has to be determined. Then it can be assessed whether these value-creation opportunities will be profitable and whether there will occur further emergent ones:

- Identify cross-selling opportunities for existing products;
- Identify opportunities/needs for new products or product updates;
- Identify opportunities/needs for new product platforms;
- Revise product planning;
- Identify technology leveraging opportunities;
- Identify opportunities/needs for new product or process technologies;
- Revise technology planning;
- Identify ideas for new markets;
- Identify opportunities to transfer technologies;
- Identify opportunities to substitute technologies;
- Identify opportunities to fuse technologies;
- Identify opportunities to integrate technologies;

- Identify opportunities to reconfigure technologies;
- Identify opportunities to add new technologies;
- Identify opportunities to eliminate technologies;
- Identify opportunities to bundle technologies.

B.4 Assessment and Selection of Technology-based Value-creation Opportunities

In order to limit the scope of the investigations and future strategy the companies have to focus on the most attractive technology-based value-creation opportunities. In order to select and assess them the following questions can be posed.

Strategic fit of value-creation opportunities

- Fit with the overall corporate and business unit strategy (fosters business objectives, etc.);
- Fit with the technology and innovation strategy (in line with the technology portfolio and core competencies, etc.);
- Fit with the competitive and development strategic plans (contributes in the short and long run to the strategic objectives, etc.);
- Reinforce the other value-creation opportunities and further integration activities (technological synergies, etc.);
- Reinforce the daily business (support sustaining value);
- Optimize the effort within the company organization and processes (can be managed with the processes, etc.);
- Optimize the efforts within the company culture (matches with the company culture, etc.);
- Optimize the efforts within the dominant business logic (match to the value drivers, etc.).

Financial assessment of value-creation opportunities

- NPV and breakeven of value-creation opportunities

Risk assessment of value-creation opportunities

Internal Risks:

- Technological risks, technological complexity, etc.;
- Resource-related risks, availability of skills, process knowledge, meta-knowledge, mgmt attention, etc.;
- Acquisition and integration-specific risks, geographically dispersed, hostile environment, cultural differences, integration complexity, etc.;

External Risks:

- Market-related risks, customer reaction, supplier behaviour, etc.;
- Competition-related risks, existing and new competitors, risks of substitution, etc.;
- Other risks, political, environmental, legal issues, etc.;

Timeliness assessment of value-creation opportunities

- When is the market demand?
- Are marketing and sales on time?
- Can the manufacturing be adapted on time?
- Can the product be developed on time?
- Is the technology ready by then?
- Is the logistics ready?
- IT support ready?

B.5 Technology Due Diligence – 1 Layer: Validate Occurrence of Future Technology-based Value Creation

First of all assess the attractiveness, level of mastery and combinability of the technology base. These attributes ensure the emergence of future technology-based value-creation opportunities. Focus on the core technologies when assessing them.

Attractiveness of technology base

- Maturity (pace-maker, key or base technology);
- Functional scope (discontinuous, wide or narrow);
- Functional performance (in relation to substitutes);
- Scalability;
- Robustness;
- Conformity with standard;
- Dependence on other proprietary or costly product or process technologies;
- Risk of being substituted.

Level of mastery of the technology base

- Number of designated specialists within the field;
- Number, soundness and quality of patents within this field (IP due diligence);
- Years of experience;
- Yearly R&D investment in the field;
- Specialty of underlying infrastructure;
- Ability to operate in test conditions;
- Level of in-house competence versus external abilities.

Combinability of the technology base

- Are the technologies sufficiently similar/redundant or complementary/supplementary/distinct?
- Is it possible to integrate the technologies in the long run due to their tacitness, context dependence, dispersion or e.g. non-modular structure?
- Are the technologies too specific, rare, or difficult to imitate to transfer them?

B.6 Technology Due Diligence – 2 Layer: Validate Profitability of Planned Technology-based Value-creation Opportunities

In the next step the potential profitability of the joint technology roadmap or pipeline has to be investigated. Thus the planned and identified value-creation opportunities are assessed.

Congruence of value-creation opportunities

The technology-based value-creation opportunities have to:

- Represent a missing link between customer needs expressed in functional requirements and functional scope provided by the linked technologies;
- Balance the market pull and technology push;
- Have a modular, complexity-reducing and elegant basic underlying (product) concept.

Relatedness of value-creation opportunities

The main questions to ask regarding the relatedness are:

- Are we acquainted with the customers and markets which we address with the value-creation opportunity?
- Have we ever managed a similar product concept before?
- Do we have sufficient experience within the technology platform?
- Is the way to develop or service the product similar to our way of doing business?

Feasibility of value-creation opportunities

The realization of the value-creation opportunities has sufficient:

- Specialists, people knowledgeable within the area;
- A leadership person managing the realization of the value creation opportunity;
- Sufficient financial resources;
- Sufficient time;
- The appropriate infrastructure;
- The required manufacturing capacity;
- Slack to remain flexible.

B.7 Technology Due Diligence – 3 Layer: Validate Potential for Successful Realization of Technology-based Value-creation Opportunities

Last but not least, it has to be assessed whether the two companies together have the ability to jointly realize the value-creation opportunities. The following questions should be posed:

Corporate culture

- Are both companies' openness conducive to technology-based value creation and are they sufficiently similar?
- Are both companies' leadership styles conducive to technology-based value creation and are they sufficiently similar?
- Are both companies' time management conducive to technology-based value creation and are they sufficiently similar?
- Are both companies' reactions to change conducive to technology-based value creation and are they sufficiently similar?
- Are both companies' diversities conducive to technology-based value creation and are they sufficiently similar?

Organizational context

- Are both companies' configurations conducive to technology-based value creation and are they sufficiently similar?
- Are both companies' orientations conducive to technology-based value creation and are they sufficiently similar?
- Are both companies' boundaries conducive to technology-based value creation and are they sufficiently similar?
- Are both companies' level of autonomy conducive to technology-based value creation and are they sufficiently similar?
- Are both companies' level of formalization conducive to technology-based value creation and are they sufficiently similar?

Dominant business logic

- Are both companies' technology strategies conducive to technology-based value creation and are they sufficiently similar?
- Are both companies' value drivers conducive to technology-based value creation and are they sufficiently similar?
- Are both companies' technology applications conducive to technology-based value creation and are they sufficiently similar?
- Are both companies' technology marketing strategies conducive to technology-based value creation and are they sufficiently similar?
- Are both companies' innovation rates conducive to technology-based value creation and are they sufficiently similar?

Notes

1 Introduction

1. A recent Boston Consulting Group Study, published in December 2003, reports that the improvements of operating cash flows and profitable investments, such as acquisitions, have increased the fundamental value of several companies. These elementary strategic actions were rewarded by the market as the market value of these companies has converged with the fundamental value and outperformed the evaluations of their competitors (see also *NZZ, Neue Zürcher Zeitung*, 2004).
2. Innovation driven industries, within this book, are understood as industries where the competitiveness of its players is significantly determined by their innovativeness. Savioz (2002: 17) has made an attempt to distinguish between high-technology or technology-based industries and low-technology industries. He discusses several input- and output-based measures such as R&D expenditures (for example: OECD has set the limit at 3.5 per cent) or innovation rate and came to the conclusion that both methodologies do not fully capture the essence of the industry dynamics. Thus the definition in this book focuses on the external demand for innovativeness and does not consider the company specific reaction to it in terms of R&D investments. This is especially appropriate within this book as technology grafting acquisitions are a means of R&D input however not calculated in this way. This definition of innovation-driven industries could eventually be further detailed by characterizing them as industries, where the participants' price/earning ratio correlates with their innovativeness. A similar study was conducted by ADL (1986–96) which showed the correlation between the average annual shareholder return and the innovativeness of a sample of 600 US companies between 1986 and 96.
3. February 26, 2003, CNET News.com Mr Tolliver, executive vice president of marketing and strategy at Sun Microsystems stated at the two-day annual analyst conference in spring this year that: 'Sun balances acquisition and innovation'. He tried to make clear that 'Innovation pays. Yet many of the company's biggest plans today came from innovation outside of Sun.' He continues: 'there are some terrific M&A (mergers and acquisitions) opportunities out there. We will continue to do that in the coming year.'
4. See Link (1988), Granstrand *et al.* (1992).
5. See also Kwrak (2002) and Sikora (2000). Furthermore, the *NZZ* (*Neue Zürcher Zeitung*, 2003) on 15 July 2003 also reported an increase in acquisitions in high-tech industries, stating the most recent examples such as Oracle's bid for Peoplesoft, the $1.3 bn acquisition of Legato Systems by its rival EMC, market leader in the data storage segment and Yahoo's $1.6 bn bid for the internet company Oventure.
6. This definition of technology-based value creation will be used throughout this book.
7. Value creation is explained by Rappaport (1999: 60) as a process whereby value is created if the rate of return of an investment is higher than the cost of capital demanded by the capital market. Value creation is calculated as the sum of discounted cash flows over a determined period of time.

258 Notes

8 Mainly described by Schumpeter (1934), Abernathy and Utterback (1978) Wernerfelt (1984), Prahalad and Hamel (1990), D'Aveni (1994), Teece *et al.* (1997), Tushman and Anderson (1997), Tschirky (1998a) and Zahra (1999).
9 There are an increasing number of authors stressing the need no only to explore the resource base via innovations but also to efficiently exploit and use the investments (Harris, Insigna, Morone, & Werle, 1996).
10 Within this book the terms capability and competence are used interchangeably.
11 For a more detailed analysis on the success and failure rate in corporate acquisitions from various perspectives see Appendix A.
12 See Lubatkin and O'Neill (1987), Singh and Montgomery (1987), Ravenscraft and Scherer (1989) and Seth (1990).
13 For further information on product championing see Burgelmann *et al.* (2001: 694).
14 This lack of a holistic approach towards acquisition management was addressed by Bower and David (2002) at the M&A Summit in 2002 in Calgary, Canada.
15 Event studies are an analysis of the impact of acquisitions on the share price of the acquirer and target within a short time period around the announcement of the acquisition. Also see Bower and David (2002) and Halpern (1983).
16 Jansen (2002) has shown that learning from acquisition integration is rarely conducted or achieved.
17 Due Diligence is the very detailed analysis of the target company. Defined in a very narrow sense it aims at validating the assumptions made in the acquisition strategy, identifying deal breakers and providing sufficient information for the calculation of the company value.
18 The acquisition and integration management process typically consists of three phases, the pre-transaction phase, the transaction phase and the post-transaction phase.
19 Berens and Strauch (2002) have shown that in only 57.8 per cent of the acquisitions analysed was a production or technical due diligence conducted. Furthermore, only 6.6 per cent of all man-hours during the due diligence were assigned to the production or technical due diligence.
20 The scope of this research is confined to strategic acquisitions in innovation-driven industries, as the term acquisition is attributed to various types of transactions in the market for corporate control, for example management-buy-out (MBO), reverse takeovers, etc. Thus it seems appropriate to focus the scope of this book more precisely on strategic acquisitions. These acquisitions are understood as one strategic path to achieve a business strategic goal such as technology leadership or market dominance and as a means to improve the operational cash flow of the firms.

The focus on innovation-driven industries can be explained by the fact that technology-based value creation is only crucial for competitiveness of a company in highly competitive and innovation-driven environments. Even though the solutions given in this book might also apply to acquisitions in other industries, the associated benefits will not have the most important impact on the competitiveness of the company.
21 Ulrich (2001: 212) describes the differences between theoretical and applied research referring to the source of problems, the type of problems, the research objective, the pursued statements, research regulation and the criteria for success.
22 See 'A case study is an empirical inquiry that investigates a contemporary phenomenon within its real-life context, especially when the boundaries between phenomenon and context are not clearly evident' Yin (1994: 13).

2 Technology, Innovation and their Management

1. See also chapter 1.1.
2. See Rappaport (1999: 60).
3. See Grant and Baden-Fuller (1995), Teece et al. (1997), Nonaka and Takeuchi (1995) and Levitt and March (1990).
4. The extent to which knowledge is or is not codifiable.
5. The extent to which knowledge is highly contextualized and co-dependent on unidentified aspects of the local environment.
6. The extent to which the resource is concentrated in the head of an individual or spread out across the minds of many.
7. See Dosi et al. (1992) and Leonard-Barton (1992).
8. For technology fusion and innovation see Kodama (1992). He describes technology fusion as the creation of new technologies from combining two existing ones; one example for technology fusion is 'mechatronics', where mechanics and electronics were fused.
9. For technology integration and innovation see also Iansiti (1998).
10. 'To produce means to combine materials and forces within our reach ... To produce other things ... means to combine these materials and forces differently' (Schumpeter, 1934: 65).
11. See Henderson and Cockburn (1994: 67).
12. Grant defines the organizational capability as the integration of individuals' specialized knowledge.
13. Kogut and Zander (1992: 391) define combinative capabilities as the ability: 'to generate new combinations of existing knowledge' and 'to exploit its knowledge of the unexplored potential of the technology'.
14. See also Hauschildt (1993b: 3).
15. See also Brockhoff (1997: 51).
16. See Tschirky (1998b: 194) for a detailed discussion of various approaches.
17. Tschirky (2000: 418) extends the integrated view of technology management to 'Enterprise Science' in order to establish the correspondence between theory and reality in technology-based companies.
18. Wolfrum (1991:72) defines the purpose of technology strategies as follows: 'Which technology from which source should be used when and on which performance level and for what purpose?'
19. The concept of market pull and technology pushed is also discussed by Brockhoff (1997: 43).
20. See Brockhoff (1997).
21. See Gerpott (1999).
22. See Bucher (2003).
23. The main representatives of this school are exponents from the Harvard Business School, such as Andrews (1971), Christensen (1973).
24. The same line of argumentation was followed by Hunt et al. (2003).
25. For a comprehensive description of various technology assessment methods see Smith et al. (1994), Megantz (1996), Razgaitis (1999), Damodaran (2001), Völker and Kasper (2002) and Hunt et al. (2003).
26. Dominant design see Utterback (1994) and Hall (2002).
27. See also technology value analysis by Tschirky (1998c: 317).
28. For a detailed discussion on the core competence evaluation see Völker and Kasper (2002).

29 See also Ashton and Klavans, (1997), Lang (1998), Reger *et al.* (1998), Lichtenthaler (2000) and Savioz (2002).
30 A good overview is given by Mitterdorfer-Schaad (2001: 42).

3 Introduction to Corporate Acquisitions

1 See for example Berens and Brauner (1999).
2 See Holzapfel and Pöllath (1997).
3 See Semler (1996: 482) and Holzapfel and Pöllath (1997: 71f).
4 See Holzapfel and Pöllath (1997: 71f).
5 See Berens and Brauner (1999: 25).
6 The terms marketing, or technology, concentric acquisitions were introduced by Sautter (1988: 22). This has become necessary after Penrose (1959: 109) has shown that, besides the dimension of products and markets, the technology dimension is required to fully describe diversification.
7 Rappaport (1986) and Gomez and Weber (1989) have pointed out the importance of value creation and the increase in shareholder value as the primary objective of all strategic actions and more particularly of all corporate acquisitions. Value creation is also defined in Chapter 1.1.
8 See Jansen (2000: 96) and Gomez (2000: 41ff).

4 Case Studies from Reality: Technology-based Value Creation in Real-life Acquisitions

1 See Haspeslagh and Jemison (1991).
2 See Chapter 1.1.
3 At this point it has to be added that acquisitions which aim to acquire a company in order to prevent others from buying this company are also excluded from this research. Even though it can be argued that these acquisitions are strategic, the associated value creation resulting from pursuing this path does not directly affect the operating cash flow of the acquirer. These acquisitions, due to their highly diverse strategic focus, do not fit into the theoretical model of this research and are thus excluded.
4 At this point the author would again like to thank the interview partners for the time devoted to the research and the detailed explanations provided.
5 This is in line with the case study approach discussed by Yin (1994) and Eisenhardt (1989a).
6 Referring to the media release on 28 January 2003 and interviews with the corporate M&A team.
7 Krones was the clear leader in filling machines with a market share of around 25 per cent (Zedtwitz, 2002).
8 Sidel was the clear market leader in blow-moulding machines with a 49 per cent market share (Zedtwitz, 2002).
9 Aseptofill had only 400 employees, whereas Fillpack had about 7000 in 1999.
10 Referring to the market annual growth of PET of 10 per cent since 1990.
11 See NZZ 26.11.2002.
12 A batch machine can treat only one wafer at a time.
13 See Bader (2000).
14 See also the internal magazine of Unaxis: Chip Unaxis (Kötter, 2002).

15 Derived from internal documents on the *Unified Platform* (Krämer, 2001).
16 See press release on the new COO (release, 2001).
17 CAM stands for Computer-Aided-Manufacturing.
18 Internal document on the project Allegra and the platform strategy.

5 Model of Reality: A New Understanding of Technology-based Value Creation in Corporate Acquisitions

1 See Hilti case.
2 Little organizational integration retains the context of the target company and thus keeps especially smaller targets innovative and supports the retention of key engineers especially within the first phases of the integration.
3 The innovativeness is dependent on the retention of key engineers.
4 This is in accordance with the observations made by Puranam *et al.* (2003).
5 Mostly after about two years.
6 This need for communication and integration for learning and the transfer of tacit knowledge was elaborated by Nonaka and his colleagues within their SECI model (Nonaka, 1991: 89f).
7 See Bower (2001).
8 See Bower and David (2002).
9 See Larrson (1993), Larsson and Finkelstein (1999) and Breggren (2001).
10 See Hitt *et al.* (1998b) and Chaudhuri and Tabrizi (1999).
11 See Hitt *et al.* (1991; 1998a; 1998b).
12 See Chakrabarti *et al.* (1994), Gerpott (1995), Bresman *et al.* (1999), Hagedoorn and Duysters (2002) and Kwak (2002).
13 The corporate context is referred to as the corporate culture, dominant business logic and organizational characteristics of a corporation.
14 Within literature this technology-based value creation potential was referred to as technology fit. However, as this term first is quite narrowly defined and secondly would eventually confuse the reader, the broader and better-defined new term of 'technology-based value creation potential' is used.
15 See Ahuja and Katila (2001) and Kwak (2002).
16 See also Tschirky (1998c) and Binder and Kanowsky (1996).
17 Often literature on technology aspects in acquisitions or on the technology due diligence focus on tools and methods to consider and analyse IP-related issues within acquisitions. Within this book the IP issues are an important aspect indicating the level of mastering a technology internally and the profoundness of technology-based value-creation opportunities; however, the technology-related scope of this book extends the boundaries of patent-related aspects.
18 See Bresman *et al.* (1999) and Ranft and Lord (2002).
19 See Hagedoorn and Duysters (2002), who state that a wide dispersion of the technology base is conducive to achieving technological synergies. This, however, stands in contrast to the finding that geographic proximity enables technology-based value creation (see Chaudhuri and Trabrizi, 1999).
20 See Hagedoorn and Duysters (2002).
21 See Kwak (2002).
22 See Hauschildt (1997), Brockhoff (1999), Mitterdorfer-Schaad (2001).
23 Several authors also describe the importance of the innovation process and or the innovation barriers. These issues relevant for innovation are considered within the initial conditions related to the contextual fit.

24 This claim can be fulfilled by separating technology and innovation projects. Thus it can be ensured that technology projects have to be finalized before the innovation project, integrating the technology, begins.
25 See Utterback *et al.* (1976).
26 Hitt *et al.* (1998b); Ernst and Vitt (2000), Hagedoorn and Duysters (2002) and Gerpott (1995) have even found a negative relation. In contrast acquisition experience seems to positively impact on general value creation in acquisition.
27 See also James *et al.* (1998).
28 For a detailed discussion of the absorptive capacity in cooperations or acquisitions see Cohen and Levinthal (1990) Zander and Kogut (1995), Lane and Lubatkin (1998) or Zahra and George (2002).
29 On the importance of the absorptive capacity on technology-intensive acquisitions see also Guinan and Greenberg (2002) or Zahra and George (2002).
30 Jemison was so far the only one to mention it (Jemison, 1988).
31 Corporate culture is understood according to Schein (1992: 12) as: 'a pattern of shared basic assumptions that the group learned as it solved its problems of external adaptation and internal integration, that has worked well enough to be considered valid and, therefore, to be taught to new members as the correct way to perceive, think, and feel in relation to those problems'.
32 These attributes were collected from researchers such as Ulrich and Probst (1988), Schein (1992) and Bleicher (1999).
33 See also (Hagedoorn & Duysters, 2002: 79).
34 As happened in the Aseptofill case.
35 See Foster and Kantrow (1988), Chakrabarti *et al.* (1994), Chaudhuri and Tabrizi (1999) and Ernst and Vitt (2000).
36 For example, Chakrabarti *et al.* (1994) proved that the efficiency of the acquisition process increases the potential for increased R&D performance after acquisitions.
37 See Chakrabarti and Souder (1987), Sen and Rubenstein (1990), Durrani *et al.* (1998), James *et al.* (1998) Chaudhuri and Tabrizi (1999).
38 There is a fundamental difference between 'systemic' and 'systematic'. 'Systemic' is a holistic understanding of elements which are related one to the other. 'Systematic', in turn, describes the procedure of action. See Daenzer (1976: 11) referred to in Savioz (2002: 38).
39 Gatekeepers are introduced by Allen (1977).
40 See James *et al.* (1998).
41 See Guinan and Greenberg (2002).
42 Similarly Alp *et al.* (1997) and Guinan and Greenberg (2002) have outlined the important role of middle managers in enabling technology-based value creation.
43 See Hitt *et al.* (1991b), James *et al.* (1998) and Devine and Lammiman (2000).
44 Kozin and Young (1994: 24) state: "Due diligence traditionally focuses on aspects such as the target's revenue and cost structures, the search for contingent liabilities, and various legal issues. The most sophisticated acquirers must go beyond these fundamentals to identify the target's desirable core competencies, carefully analyze and value them and creatively structure purchase-and-sale agreements to ensure that the skills and knowledge that comprise the core competencies are secured."
45 See Hitt *et al.* (1998b) Foster and Kantrow (1988), Granstand and Sjölander (1990), Chaudhuri and Tabrizi (1999), Bryer and Simensky (2002), Slowinski *et al.* (2002), Cullinan *et al.* (2004).
46 See Foster and Kantrow (1988) and Chaudhuri and Tabrizi (1999).
47 See Granstand and Sjölander (1990), Kozin and Young (1994), Bryer and Simensky (2002), Slowinski *et al.* (2002).

48 See Chakrabarti and Souder (1987).
49 See Guinan and Greenberg (2002) and Chakrabarti and Souder (1987).
50 See Granstrand and Sjölander (1990).
51 See Hardtmann (1996), Hilbert (2002), Dankl and müller (2002) and Kurr (1999).
52 On the importance of mutual respect and trust in acquisition respectively joint R&D teams see Mace and Montgomery (1962), Foster and Kantrow (1988), Jemison (1988), Shleifer and Summers (1988), Ring and Ven (1992), Birkinshaw and Hakanson (2000), Croyle and Kager (2002) and Lunnan and Barth (2003).
53 See Granstrand and Sjölander (1990) and Gerpott (1995).
54 Chakrabarti and Souder (1987), Hitt et al. (1991b) and Bresman et al. (1999), for example, outline the impact of the exchange of people.
55 For further investigations on the impact of resource deployment see Capron (1999) and on knowledge transfer see Bresman et al. (1999).
56 For a distinction between human and task integration see also Birkinshaw and Hakanson (2000).
57 See Jemison (1988) and Ranft and Lord (2002).
58 See Gerpott (1995).
59 See James et al. (1998).
60 See Granstrand and Sjölander (1990).
61 See Chaudhuri and Tabrizi (1999).
62 See Jemison (1988) and Chaudhuri and Tabrizi (1999).
63 See Devine (1984), Granstrand and Sjölander (1990), Kozin and Young (1994), James et al. (1998), Chaudhuri and Tabrizi (1999), Ranft and Lord (2002), Slowinski et al. (2002).
64 See Foster and Kantrow (1988), Devine (1984), Bresman et al. (1999) and Ranft and Lord (2002).
65 See Slowinski et al. (2002).
66 See Galunic and Rodan (1998) and Devine (1984).
67 The term corporate coherence was explained by Dosi et al. (1992), who explain that companies can reach an unstable status of corporate coherence which is described by its internal fit with its external developments and the dynamic usage of internal synergies.
68 This phenomenon was also identified by Birkinshaw (1999).
69 This was especially emphasized by the HR responsible for the Lotus Notes integration at IBM.
70 At this point most researchers mention the Not-Invented-Here (NIH) syndrome as a main hindrance to technology-based value creation in the course of the integration. Interestingly, within this research the NIH syndrome was not encountered as usually supposed. The engineers rarely hesitated to cooperate or even to take in new technologies. The only problems which occurred were at the levels of middle and higher management in the R&D area. The heads of R&D or of certain engineering groups were the ones who hindered cooperation and technology transfer, not the engineers.

6 Technology-based Strategic Acquisition and Integration Management

1 See also Mintzberg (1990) on strategy formation processes.
2 See, for example, Huemer, 1991; Reißner, 1992; Frank, 1993; World Law Group, 1995; Whalley & Heymann, 1996; Steinöcker, 1998; Rockholtz, 1999a; Jansen, S. A.,

2000b; Müller-Stewens, 2000; Picot, Nordmeyer, & Pribilla, 2000; Lucks & Meckl, 2002.

3   This argument was brought up by Barney (1988: 71) stating: 'Rather only when bidding firms enjoy private and uniquely valuable synergistic cash flow with targets, inimitable and uniquely valuable synergistic cash flow with targets, or unexpected synergistic cash flows, will acquiring a related firm result in abnormal returns for the shareholders of bidding firms.'
4   This process description intentionally excludes hostile takeovers, especially as they have been proven to be destructive of technology based value creation.
5   See Gomez and Weber (1989), Kirchner (1991), Eiffe and Mölzer (1993), Steinöcker (1998), Berens and Brauner (1999), Rockholtz (1999a), Jansen (2000b) and Bild *et al.* (2002).
6   This particularly applies to the external developments, the corporate structure and organization and the cultural characteristics of the target.
7   See the Tim Sauber's book (2003) for a process for developing innovation architecture.
8   Here the focus is on process technologies applied within the engineering, development and research activities as the other value creation activities are mostly covered by other functional teams (IT, manufacturing, and etc).
9   Similar to the quick gains and long-term success explained in 5.1.
10 See also Porter (1980).
11 The different modes of acculturation within acquisitions were discussed by Berry (1983) and Nahavandi and Malekzadeh (1988).
12 Also compare the dual responsibilities of Starrag's CEO and Head of R&D during the fist months of the integration or the established Think Tank in the case studies discussed in chapter 4.
13 Another example of temporary boundary-spanning structures occurred at ABB's acquisition of Elsag Bailey. In order to foster technology integration the worldwide engineering teams were temporarily headed by one single R&D manager who coordinated and distributed the competencies and development activities.
14 For additional information on a Letter of Intent or other types of Due Diligence see Berens and Brauner (1999).
15 See also Berens and Brauner (1999); Crilly (1993), Rockholtz (1999b) and Berens and Strauch (1999).
16 This was the case at the Unitron acquisition, where the poor technological competencies of Unitron's suppliers resulted in delays in the product launch.
17 See Klavans (1994), Rivette and Kline (2000) and Breitzmann and Thomas (2002).
18 Another interesting aspect would be to investigate the question whether there is a difference if either the acquirer or the target has a more innovation-promoting context. Within this research no conclusions could be drawn; thus the absolute value of the difference in contextual characteristics indicates the difference between target and acquirer.
19 See also Bannert and Tschirky (2004).

Appendix A – Value Creation in Corporate Acquisitions

1   Whereas only very few researchers draw implications for practitioners.
2   The concept was introduced by Manne (1965), stating that the control of corporations constitutes a valuable asset.

3 Jensen and Ruback (1983: 6) describe the market for corporate control as 'the arena in which alternative management teams compete for the right to manage corporate resources'.
4 For a detailed overview of event study methodology see Brown and Warner (1980) or Halpern (1983).
5 See Muth (1961) and Schipper and Tompson (1983).
6 An disquisition on the rationality or irrationality of investors regarding M&A was given by Bower and David (2002).
7 For an overview of agency theory see Eisenhardt (1989b). Furthermore, see Jensen and Meckling (1976), Fama (1980) and Wright *et al.* (2001).
8 See Jensen and Ruback (1983) and Jarrel *et al.* (1988).
9 See Banerjee and Owers (1992), Caper (1990), Sirower (1994), Healy *et al.* (1997) for further study overview.
10 See Jensen and Ruback (1983), Eckbo (1985) and King (2004).
11 For a detailed overview of the various impact factors and the associated research see Datta *et al.* (1992).
12 The model goes back to the mid-1960s, to William Sharpe (1964) and John Lintner (1965).
13 Huemer (1991) has shown that in 20 per cent of acquisitions tax manipulation has a positive impact.
14 See Jarrel *et al.* (1988).
15 This is also referred to as the free-cash-flow hypothesis (Jensen, 1986).
16 Roll's (1986) reason for failure: the overbearing assumptions of bidders that their valuations are correct.
17 See also Lubatkin (1983).
18 See Haspeslagh and Jemison (1991: 300).
19 See Kitching (1967), Kusewitt (1985), Fowler and Schmidt (1988), Duncan and Savill (1997), Lubatkin *et al.* (1998).
20 See Lubatkin and O'Neill (1987) and Singh and Montgomery (1987).
21 See Salter and Weinhold (1979).
22 See Lewellen (1971), Higgins and Schall (1975), Leontiades (1986).
23 See Salter and Weinhold (1979), Bettis (1981), Christensen and Montgomery (1981), Bettis and Hall (1982), Singh and Montgomery (1987), Shelton (1988), Ramaswamy (1997).
24 See Salter and Weinhold (1979), Lubatkin (1983), Seth (1990).
25 See Davis (1968).
26 Lubatkin and O'Neill (1987), Seth (1990) and Kaplan and Weisbach (1992) have not found any difference in success between related and unrelated acquisitions; however, unrelated acquisitions are four times more divested.
27 See Elgers and Clark (1980), Chatterjee (1986), Agrawal *et al.* (1992).
28 See Halbelian (1999), Gagnon and Sheu (2000), Finkelstein and Haleblian (2002), Rovit (2003).
29 See Hayes (1979), Sutton (1983), Marks and Mirvis (1985; 1998), Sales and Mirvis (1985), Buono *et al.* (1988).
30 See Walsh (1988; 1989), Hambrick and Cannella (1993), Krug and Hegarty (1997), Davis and Nair (2003).
31 See Sinetar (1981), Bastien (1987), Napier *et al.* (1989), Schweiger and DeNisi (1991), Dooley and Zimmerman (2002).
32 See Hirsch (1987) and Schneider and Dunbar (1987).
33 See Buono and Lewis (1989), Datta (1991), Mendonca and Kanungo (1994), Testa *et al.* (2003), Frank (1993) and Cartwright and Cooper (1993).

34 The definition and separation of cultural and organizational fit has not been agreed upon and thus depends on the author. Whereas some researchers differentiate between these two terms, others refer to only one concept.
35 See Bastien and Ven (1986), Jemison and Sitkin (1986a), Haspeslagh and Farquhar (1987), Haspeslagh and Jemison (1991) and Napier (1989).
36 See Duhaime and Schwenk (1985), Jemison and Sitkin (1986b), Shrivastava (1986) and Haunschild *et al.* (1994).

Bibliography

Abernathy, W. J. and Clark, K. B. (1985). Innovation: Mapping the Winds of Creative Destruction. *Research Policy* 14: 3–22
Abernathy, W. J. and Utterback, J. M. (1978). Patterns of Industrial Innovation. *Technology Review* 80(7): 41–7
Agrawal, A., Jaffe, J. F., et al. (1992). The Postmerger Performance of Acquiring Firm: A Re-examination of an Anomaly. *Journal of Finance* 47: 1605–71
Ahuja, G. and Katila, R. (2001). Technological Acquisitions and the Innovation Performance of Acquiring Firms: A Longitudinal Study. *Strategic Management Journal* 22: 197–220
Allen, T. J. (1977). *Managing the Flow of Technology: Technology Transfer and the Dissemination of Technological Information within the R&D Organization*: The Massachusetts Institute of Technology
Alp, N., Alp, B. and Omurtag, Y. (1997). The Influence of Decision Makers for New Technology Acquisition. *Computers and Industrial Engineering* 33(1–2): 3–5
Anand, J. and Singh H. (1997). Asset Redeployment, Acquisitions and Corporate Strategy in Declining Industries. *Strategic Management Journal* 18 (Summer Special Issue): 99–118
Andrews, K. R. (1971). *The Concept of Corporate Strategy*. Homewood, Ill.: Irwin
Ansoff, I. (1965). *Corporate Strategy*. New York: McGraw-Hill
Arthur D. Little (2002). Venturing Issue per Innovation. Study Report. www.adl.com/insights/studies/pdf/corporate_venturing_study_report.pdf
Ashton, W. B. and Klavans, R. A. (1997). An Introduction to Technical Intelligence in Business. In Ashton, W. B. and Klavans, R. A. (eds.), *Keeping Abreast of Science and Technology Technical Intelligence in Business*: 5–22. Columbus, Ohio: Batell Press
Bader, M. (2000). Becoming a 'One Stop'. *Layers* Issue 15: 4
Balakrishnan, S. (1988). The Prognostics of Diversifying Acquisitions. *Strategic Management Journal* 9: 185–96
Banerjee, A. and Owers, J. E. (1992). Wealth Reduction in White Knight Bids. *Financial Management* 21: 48–57
Bannert, V. and Tschirky, H. (2004). Integration Planning for Technology Intensive Acquisitions. *R&D Management* 34(5): 477–90
Barney, J. (1988). Returns to Bidding Firms in Mergers and Acquisitions: Reconsidering the Relatedness Hypothesis. *Strategic Management Journal* 9: 71–78
Barney, J. (1991). Firm Resources and Sustained Competitive Advantage. *Journal of Management* 17(1): 99–120
Bastien, D. T. (1987). Common Patterns of Behaviour and Communication in Corporate Mergers and Acquisitions. *Human Resource Management* 26: 17–33
Bastien, D. T. and Ven, A. H. V. d. (1986). *Managerial and Organizational Dynamics of Mergers and Acquisitions*. Minnesota, University of Minesota: Strategic Management Research Center
Baysinger, B. and Hoskisson, R. E. (1989). Diversification Strategy and R&D Intensity in Multiproduct Firms. *Academy of Management Journal* 32(2): 310–32
Berens, W. and Brauner, H. U. (1999). *Due Diligence bei Unternehmensakquisitionen* (Due Diligence of Corporate Acquisitions). Stuttgart: Schäffer-Poeschel

Berens, W., Hoojan, A. and Strauch, J. (1999). Planung und Durchführung der Due Diligence (Planning and Execution of Due Diligence). In Berens, W. and Brauner, H. U. (eds.), *Due Diligence bei Unternehmensakquisitionen* (Due Diligence of Corporate Acquisitions). Stuttgart: Schaeffer-Poeschel

Berens, W. and Strauch, J. (1999). Herkunft und Inhalt des Begriffes Due Diligence (Origin and Content of the teer 'Due Diligence'). In Berens, W. and Brauner, H. U. (eds.), *Due Diligence bei Unternehmensakquisitionen* (Due Diligence of Corporate Acquisitions – An Empirical Study): 3. Stuttgart: Schäffer-Poeschel

Berens, W. and Strauch, J. (2002). *Due Diligence bei Unternehmensakquisitionen – eine empirische Untersuchung*. Frankfurt am Main: Peter Lang/Europäischer Verlag der Wissenschaften

Berry, J. W. (1983). Acculturation: A Comparative Analysis of Alternative Forms. In Samuda, R. J. and Woods, S. L. (eds.), *Perspectives in Immigrant and Minority Education*: 66–77. Lanham, Md.: University Press of America

Bettis, R. A. (1981). Performance Differences in Related and Unrelated Diversified Firms. *Strategic Management Journal* 2: 379–94

Bettis, R. A. and Hall, W. K. (1982). Diversification Strategy, Accounting Determined Risk, and Accounting Determined Return. *Academy of Management Journal* 45(2): 254–64

Bild, M., Cosh, A., Guest, P. and Runsten, M. (2002). Do Takeovers Create Value: A Residual Income Approach on U.K. Data. ESRC centre for Business Research, University of Cambridge, Working Paper No. 252

Binder, V. A. and Kanowsky, J. (1996). *Technologiepoteniale* (Technology Potentials). Wiesbaden: Deutscher Universitäts-Verlag

Birkinshaw, J. (1999). Acquiring Intellect: Managing the Integration of Knowledge-Intensive Acquisitions. *Business Horizons*: 33–9

Birkinshaw, J. and Hakanson, L. (2000). Managing the Post-Acquisition Integration Process: How the Human Integration and Task Integration Processes Interact to Foster Value Creation. *Journal of Management Studies* 37(3): 395–425

Bleicher, K. (1996). *Das Konzept Integriertes Management* (The Concepts of Integrated Management): 4. New York: Campus Verlag

Bleicher, K. (1999). *Das Konzept Integriertes Management* (The Concepts of Integrated Management). Frankfurt/New York: Campus Verlag

Blex, W. and Marchal, G. (1990). Risiken im Akquisitionsprozess – ein Überblick. *BFuP* 42(2): 85–103 (Risks in the Acquisition Process – An Overview)

Bower, J. L. (2001). Not All M&As Are Alike – and That Matters. *Harvard Business Review* March: 93–101

Bower, J. L. and David, D. K. (2002). When We Study M&A, What Are We Learning? M&A Summit Conference: Calgary, Canada

Breggren, C. (2001). Mergers, MNES and Innovation – The Need for New Research Approaches. *Scandinavian Journal of Management* uncorrected proof: 19

Breitzmann, A. and Thomas, P. (2002). Using Patent Citation Analysis to Target/value M&A Candidates. *Research & Technology Management*: 28–36

Bresman, H., Birkinshaw, J. and Nobel, R. (1999). Knowledge Transfer in International Acquisitions. *Journal of International Business Studies* 30(3): 439–62

Brockhoff, K. (1997). *Forschung und Entwicklung Planung und Kontrolle* (Research and Development Planning and Controlling). München: Oldenbourg

Brockhoff, K. (1999). *Forschung und Entwicklung: Planung und Kontrolle*. München: Oldenbourg

Brodbeck, H., Birkinmeier, B. and Tschirky H. (1995). Neue Entscheidungsstrukturen des Integrierten Technologie-Managements (New Decision-making Structures of the Integrated Technolgy Management). *Die Untesnehmung* 49(2): 107–23

Brown, S. J. and Warner, J. B. (1980). Measuring Security Price Performance. *Journal of Financial Economics* 8: 205–58

Bryer, L. and Simensky, M. (2002). *Intellectual Property Assets in Mergers and Acquisitions*. New York: John Wiley & Sons

Bucher, P. (2003). Integrated Technology Roadmapping: Design and Implementation for Technology-Based Multinational Enterprises. Dissertation, ETH Zurich, Zurich

Buono, A. F., Bowditch, J. L. and J. W. Lewis, I. (1988). The Cultural Dynamics of Transformation: The Case of a Bank Merger. In Kilmann, R., Covin, T. and Associates (eds.), *Corporate Transformation: Revitalizing Organizations for a Competitive World*: 497–522. San Francisco: Jossey-Bass

Buono, A. F. and Lewis, J. W. (1989). When Cultures Collide: The Anatomy of a Merger. *Human Relations* 38: 477–500

Burgelman, R. A., Maidique, M. A. and Wheelwright, S. C. (2001). *Strategic Management of Technology and Innovation*. New York: McGraw-Hill

Capron, L. (1999). The Long-Term Performance of Horizontal Acquisitions. *Strategic Management Journal* 20: 987–1018

Capron, L. and Mitchell, W. (1998). Bilateral Resource Redeployment and Capabilities Improvement Following Horizontal Acquisitions. *Industrial and Corporate Change* 7(3): 453–84

Carper, W. (1990). Corporate Acquisitions and Shareholder Wealth: A Review and Exploratory Analysis. *Journal of Management* 16: 807–23

Cartwright, S. and Cooper, C. L. (1993). The Role of Culture Compatibility in Successful Organizational Marriage. *Academy of Management Executive* 7(2): 57–70

Chakrabarti, A., Hauschildt, J. and Süverkrüp, C. (1994). Does it Pay to Acquire Technological Firms? *R&D Management* 24(1): 47–56

Chakrabarti, A. K. and Souder, W. E. (1987). Technology, Innovation and Performance in Corporate Mergers: A Managerial Evaluation. *Technovation* 6: 103–14

Chappuis, B. E., Frick, K. A. and Roche, P. J. (2004). High-tech Mergers Take Shape. *The McKinsey Quarterly* 1: 61–71

Chatterjee, S. (1986). Types of Synergy and Economic Value: The Impact of Acquisitions on Merging and Rival Firms. *Strategic Management Journal* 7: 119–39

Chaudhuri, S. and Tabrizi, B. (1999). Capturing the Real Value in High-Tech Acquisitions. *Harvard Business Review* September–October: 123–30

Chesbrough, H. W. (2003). The Era of Open Innovation. *MIT Sloan Management Review* Spring: 35–41

Christensen, H. K. and Andrews, K. R. (1973). *Business Policy. Text and Cases*. Homewood, Ill.: Dow-Jones-Irwin

Christensen, H. K. and Montgomery, C. A. (1981). Corporate Economic Performance: Diversification Strategy versus Market Structure. *Strategic Management Journal* 2: 327–43

Clemente, M. N. and Greenspan, D. S. (1998). *Winning at Mergers and Acquisitions*. New York, Chichester, Weinheim, Brisbane, Singapore, and Toronto: John Wiley & Sons, Inc.

Cohen, W. M. and Levinthal, D. A. (1990). Absorptive Capacity: A New Perspective on Learning and Innovation. *Administrative Science Quarterly* 35(1): 128–53

Copeland, T. and Weston, F. J. (1988). *Financial Theory and Corporate Policy* (3rd edn.). The Addison-Wesley Series in Finance. Menlo Park (CA): Addison-Wesley

Crilly, W. M. (1993). *Due Diligence Handbook*. Newport: Pacific Associates

Croyle, R. and Kager, P. (2002). Giving Mergers a Head Start. *Harvard Business Review*: 20–1

Cullinan, G., Roux, J.-M. L. and Weddingen, R.-M. (2004). When to Walk Away from a Deal. *Harvard Business Review* April: 9

Daenzer, W. F. (ed.)(1976). *System Engineering*. Zurich: Verlag Industrielle Organisation
Damodaran, A. (2001). *The Dark Side of Valuation*. London: Financial Times, Prentice Hall
Dankl, A. and Müller, M. (2002). Technical Due Diligence. In Kranebitter, G. (ed.), *Due Diligence: Risikoanalyse im Zuge von Unternehmenstransaktionen*: 135–44. München: Redline Wirtschaft bei verlag moderne industrie
Datta, D. K. (1991). Organizational Fit and Acquisition Performance: Effects of Post-acquisition Integration. *Strategic Management Journal* 12: 281–97
Datta, D. K., Pinches, G. E. and Narayanan, V. K. (1992). Factors Influencing Wealth Creation from Mergers and Acquisitions: A Meta-analysis. *Strategic Management Journal* 13: 67–84
D'Aveni, R. (1994). *Hypercompetition. Managing the Dynamics of Strategic Manoeuvring*. New York: Free Press
Davenport, T. and Prusak, L. (1998). *Working Knowledge: How Organizations Manage What They Know*. Cambridge, MA: Harvard Business School Press
Davis, R. and Nair, A. (2003). A Note on Top Management Turnover in International Acquisitions. *Management International Review* 43(2): 171–83
Davis, R. E. (1968). Compatibility in Corporate Marriages. *Harvard Business Review*: 86–93
Devine, I. (1984). Organizational Adaptation to Crisis Conditions and Effects on Organizational Members. *Academy of Management Proceedings*: 163–67
Devine, M. and Lammiman, J. (2000). Original Synergy. *People Management* 13: 28–33
Dooley, K. J. and Zimmerman, B. J. (2002). Merger as Marriage: Communication Issues in Post-merger Integration, M&A Summit 2002: Calgary, Canada
Dosi, G., Rumelt, R., Teece, D. J. and Winter, S. G. (1992). *Towards a Theory of Corporate Coherence: Preliminary Remarks*. University of Rome, University of California at Los Angeles, University of California at Berkeley
Duhaime, I. M. and Schwenk, J. H. (1985). Conjectures on Cognitive Simplification in Acquisition and Divestment Decision Making. *Academy of Management Review* 10: 287–95
Duncan, R. and Savill, B. (1997). Strategic Perspective on Euoprean Cross-border Acquisitions: A View from Top European Executives. *European Management Journal* 16(2): 124–35
Durrani, T. S., Forbes, S. M., Broadfoot, C. and Carrie, A. S. (1998). Managing the Technology Acquisition Process. *Technovation* 18(8/9): 523–28
Eckbo, B. E. (1985). Mergers and the Market Concentration Doctrine: Evidence from the Capital Market. *Journal of Business* 58(3): 325–49
Eiffe, F. F. and Mölzer, W. (1993). *Mergers & Acquisitions – Leitfaden zum Kauf und Verkauf von Unternemen* (Guidance for the Acquisition and Sale of Corporations). Wien: Service Fachverlag
Eisenhardt, K. M. (1989a). Building Theories from Case Study Research. *Academy of Management Review* 14(4): 532–50
Eisenhardt, K. M. (1989b). Agency Theory: An Assessment and Review. *Strategic Management Journal* 14: 57–74
Eisenhardt, K. M. (1999). Strategy as Strategic Decision Making. *Sloan Management Review*: 65–72
Elgers, P. T. and Clark, J. J. (1980). Merger Types and Stockholder Returns: Additional Evidence. *Financial Management* 9: 66–72
Ernst, H. and Vitt, J. (2000). The Influence of Corporate Acquisitions on the Behaviour of Key inventors. *R&D Management* 30(2): 105–20
Fama, E. F. (1980). Agency Problems and the Theory of the Firm. *Journal of Political Economy* 88: 288–307

Finkelstein, S. (1986). *The Acquisition Integration Process*, Working Paper: 4. Graduate School of Business, Columbia University

Finkelstein, S. and Haleblian, J. (2002). Understanding Acquisition Performance: The Role of Transfer Effects. *Organization Science* 13(1): 36–47

Foster, R. N. (1986). *Innovation, The Attacker's Advantage*. New York: Wiley

Foster, R. N. and Kantrow, A. M. (1988). Making Post-merger R&D Effective. *Research-Technology Management* 31: 47–51

Fowler, K. and Schmidt, D. R. (1988). Tender Offers, Acquisitions and Subsequent Performance in Manufacturing Firms. *Academy of Management Journal* 31(4): 962–74

Frank, G.-M. (1993). *Probleme und Erfolgsfaktoren bei der Übernahme von Unternehmen* (Problems and Success Factor of the Acquisition of Companies). Stuttgart: Schäffer-Poeschel

Gagnon, R. J. and Sheu, C. (2000). The Impact of Learning, Forgetting and Capacity Profiles on the Acquisition of Advanced Technology. *The International Journal of Management Science* 28: 51–76

Galunic, D. C. and Eisenhardt, K. M. (2001). Architectural Innovation and Modular Corporate Forms. *Academy of Management Journal* 44(6): 1229–49

Galunic, D. C. and Rodan, S. (1998). Resource Recombinations in the Firm: Knowledge Structures and the Potential for Schumpeterian Innovation. *Strategic Management Journal* 19(12): 1193–201

Gerpott, T. J. (1995). Successful Integration of R&D Functions after Acquisition: An Exploratory Empirical Study. *R&D Management* 25: 161–79

Gerpott, T. J. (1999). *Strategisches Technologie- und Innovationsmanagement* (Strategic Technology and Innovation Management). Stuttgart: Schäffer-Poeschel Verlag

Gomez, P. (2000). Management des Unternehmens-Portfolios – Wertsteigerung durch Akquisition (Management of the Company Portfolio – value Creation through Acquisitions). In Picot, A., Nordmeyer, A. and Pribilla, P. (ed.), *Management von Akquisitionen*: 21–39. Stuttgart: Schäffer-Poeschel Verlag

Gomez, P. and Weber, B. (1989). *Akquisitionsstrategie: Wertsteigerung durch die Übernahme von Unternehmungen* (Acquisition Strategy: Value Creation through the Acquistion of Corporation). Stuttgart: Schäffer

Granstrand, O., Bohlin, E., Oskarsson, C. and Sjöberg, N. (1992). External Technology Acquisition in Large Multi-technology Corporations. *R&D Management* 22(2): 111–33

Granstrand, O. and Sjölander, S. (1990). The Acquisition of Technology and Small Firms by Large Firms. *Journal of Economic Behavior and Organization* 13: 367–86

Grant, R. M. (1996). Prospering in Dynamically-Competitive Environments: Organizational Capability as Knowledge Integration. *Organization Science* 7(4): 375–87

Grant, R. M., and Baden-Fuller, C. (1995). A Knowledge-Based Theory of Inter-firm Collaboration. *Academy of Management Best Paper Proceedings*: 17–21

Greengard, S. (1999). Due Diligence: The Devil in the Details. *Workforce*: 68–74

Guinan, P. J. and Greenberg, D. (2002). Strategies for Leveraging Knowledge and Creating Value from M&As in Technology-Based Industries, M&A Summit 2002: 1–13: Calgary, Canada

Hagedoorn, J. and Duysters, G. (2002). The Effect of Mergers and Acquisitions on the Technological Performance of Companies in a High-Tech Environment. *Technology Analysis & Strategic Management* 14(1): 67–85

Haleblian, J. (1999). The Influence of Organizational Acquisition Experience on Acquisition Performance: A Behavioral Learning Perspective. *Administrative Science Quarterly* 44: 29–56

Hall, B. (1988). The Effect of Takeover Activity on Corporate Research and Development. In Auerbach, A. J. (ed.), *Corporate Takeovers: Causes and Consequences*. Chicago and London: University of Chicago Press

Hall, K. (2002). *Ganzheitliche Technologiebewertung. Ein Modell zur Bewertung unterschiedlicher Produktionstechnologien* (Holistic Technology Assessment. A Model to Assess Different Production Technologies) (Deutscher Universitäts-Verlag ed.). Wiesbaden: Deutscher Universitäts-Verlag

Halpern, P. J. (1983). Corporate acquisitions: A Theory of Special Case? A Review of Event Studies Applied to Acquisitions. *Journal of Finance* 38: 297–317

Hambrick, D. C. and Cannella, A. A. (1993). Relative Standing: A Framework for Understanding Departures of Acquired Executives. *Academy of Management Review* 8: 398–405

Hardtmann, G. (1996). *Die Wertsteigerungsanalyse im Managementprozess* (The Value Creation Analysis within the Arrangement Process). Wiesbaden

Harris, R. C., Insigna, R. C., Morone, J. and Werle, M. J. (1996). The Virtual R&D Laboratory. *Research & Technology Management* 39: 32–36

Harrison, J. S., Hitt, M., Hoskisson, R. E. and Ireland, R. D. (1991). Synergies and Post-Acquisition Performance: Differences versus Similarities in Resource Allocations. *Journal of Management* 17(1): 173–90

Haspeslagh, P. C. and Farquhar, A. (1987). The Acquisition Integration Process: Contingent Framework, Seventh Annual International Converence of the Strategic Management Society: Boston

Haspeslagh, P. and Jemison, D. B. (1991). *Managing Acquisitions: Creating Value through Corporate Renewal*. New York: The Free Press, Simon & Schuster Inc.

Haunschild, P. R., Davis-Blake, A. and et al. (1994). Managerial Overcommitment in Corporate Acquisition Process. *Organization Science* 5: 528–40

Hauschildt, J. (1993a). Determinanten des Innovationserfolges (Determinants of the Innovation Success). In Hauschildt, J. and Grün, O. (eds.), *Ergebnisse empirischer Betriebswissenschaftlicher Forschung* (Empirical Results of Management Research): 295–326. Stuttgart: Schaeffer-Poeschel Verlag

Hauschildt, J. (1993b). Innovationsmanagement (Innovation Management). In Frese, E. (ed.), *Handwörterbuch der Organisation* (Dictionary of the Organisation), vol. 3: 1029–41. Stuttgart: Poeschel

Hauschildt, J. (1997). *Innovationsmanagement* (Innovation Management). München: Franz Vahlen

Hayes, R. H. (1979). The Human Side of Acquisitions. *The Management Review* 68(11): 41–6

Healy, P. M. and Krishna, G. (1997). Which Takeovers Are Profitable? Strategic or Financial. *Sloan Management Review* 38(4): 45–57

Henderson, R. M. and Clark, K. B. (1990). Architectural Innovation: The Reconfiguration of Existing Product Technologies and the Failure of Established Firms. *Administrative Science Quarterly* 35: 9–30

Henderson, R. and Cockburn, I. (1994). Architectural Innovation: The Reconfiguration of Existing Product Technologies and the Failure of Established Firms. *Administrative Science Quarterly* 35: 9–30

Higgins, R. C. and Schall, L. D. (1975). Corporate Bankruptcy and Conglomerate Merger. *Journal of Finance* 30(1): 93–113

Hilbert, D. A. (2002). Technische Due Diligence bei Unternehmensakquisitionen (Technology Due Diligence of Corporate Acquisitions). *M&A Review* 7: 376–80

Hirsch, P. M. (1987). *Pack Your Own Parachute*. Reading, MA: Addison-Wesley

Hitt, M., Hoskisson, R. E. and Ireland, R. D. (1990). Mergers and Acquisitions and Managerial Commitment to Innovation in M-Form Firms. *Strategic Management Journal* 11: 29–47

Hitt, M., Hoskisson, R. E., Ireland, R. D. and Harrison, J. S. (1991a). Effects of Acquisitions on R&D Inputs and Outputs. *Academy of Management Journal* 34(3): 693–706

Hitt, M., Hoskisson, R. E., Ireland, R. D. and Harrison, J. S. (1991b). Are Acquisitions a Poison Pill for Innovation? *Academy of Management Executive* 5(4): 22–34

Hitt, M., Harrison, J., Ireland, R. D. and Best, A. (1998a). Attributes of Successful and Unsuccessful Acquisitions of US Firms. *British Journal of Management* 9: 91–114

Hitt, M. A., Hoskisson, R. E., Duane, R. and Harrison, J. (1998b). Acquisitive Growth Strategy and Relative R&D Intensity: The Effects of Leverage, Diversification, and Size. *Academy of Management Proceedings*: 22–6

Holzapfel, H.-J. and Pöllath, R. (1997). *Unternehmenskauf in Recht und Praxis: rechtliche und steuerliche Aspekte* (Acquisition in Law and Practice: Legal and Tax Aspects). Köln: Verlag Kommunikationsforum Recht, Wirtschaft, Steuern

Hoskisson, R. E. and Hitt, M. A. (1988). Strategic Control Systems and Relative R&D Investment in Large Multiproduct Firms. *Strategic Management Journal* 9: 605–21

Huemer, F. (1991). *Mergers & Acquisitions: Strategische und finanzielle Analyse von Unternehmensübernahmen* (Strategic and Financial Analysis of Corporate Acquisitions). Frankfurt am Main: Peter Lang

Hunt, F. H., Probert, D. R., Wong, J. C. and Phaal, R. (2003). Valuation of Technology: Exploring a Practical Hybrid Model, *Picmet: Technology Management for Reshaping the World*: Portland

Iansiti, M. (1998). Technology Integration: Matching Technology and Context. In Dorf, R. C. (ed.), *The Technology Management Handbook*: 3–60: Boca Raton, FLA: CRC Press and Springer Verlag

James, A. D., Georghiou, L. and Metcalfe, J. S. (1998). Integrating Technology into Merger and Acquisition Decision Making. *Technovation* 18(8/9): 563–73

Jansen, S. (2000a). Szenen einiger Unternehmens-Ehen: Vier Hochzeiten und drei Todesfälle (Scenes of Some Corporate Marriages: Four Weddings and Three Deaths). *Frankfurter Allgemeine Zeitung*

Jansen, S. A. (2000b). *Mergers & Acquisitions: Unternehmensakquisition und -kooperation* (Corporate Acquisitions and Cooperations). Wiesbaden: Betriebswirtschaftlicher Verlag Dr. Th. Gabler

Jansen, S. A. (2002). Kontrolle von Unternehmenszusammenschlüssen (Control of Corporate Acquisitions). *Mergers & Acquisitions* 5: 265–71

Jarrel, G. A., Bricklea, J. A. and Netter, J. M. (1988). The Market for Corporate Control: The Empirical Evidence Since 1980. *Journal of Economic Perspectives* 2: 21–48

Jemison, D. (1988). Value Creation and Acquisition Integration: The Role of Strategic Capability Transfer, *Corporate Reorganization Through Mergers, Acquisitions, and Leveraged Buyouts*, Supplement 1: 191–218: JAI Press Series: Advances in the Study of Entrepreneurship, Innovation and Economic Growth.

Jemison, D. B. and Sitkin, S. B. (1986a). Corporate Acquisitions: A Process Perspective. *Academy of Management Review* 11(1): 145–63

Jemison, D. B. and Sitkin, S. B. (1986b). Acquisition: The Process Can Be a Problem. *Harvard Business Review*: 107–16

Jensen, M. and Meckling, W. (1976). Theory of the Firm: Managerial Behavior, Agency Costs, and Capital Structure. *Journal of Financial Economics* 3: 363

Jensen, M. C. (1986). Agency Cost of Free Cash Flow, Corporate Finance, and Takover. *AER* 76: 323–9

Jensen, M. C. and Ruback, R. S. (1983). The Market for Corporate Control – The Scientific Evidence. *Journal of Financial Economics* 11: 3–50

Kaplan, S. and Weisbach, M. S. (1992). The Success of Acquisitions: Evidence from Divestitures. *Journal of Finance* 67(1): 107–38

King, D. R., Dalton, D. R., Daily, C. M. and Covin, J. G. (2004). Meta-Analyses of Post-Acquisition Performance: Indications of Unidentified Moderators. *Strategic Management Journal* 25: 187–200

Kirchner, M. (1991). *Strategisches Akquisitionsmanagement im Konzern* (Strategic Acquisition Management within the Corporation). Wiesbaden: Gabler

Kitching, J. (1967). Why Do Mergers Miscarry? *Harvard Business Review* 45: 84–101

Klavans, R. (1994). The Measurement of a Competitor's Core Competence. In Hamel, G. and Heene, A. (eds.), *Competence-Based Competition*. Chichester: Wiley & Sons Ltd

Kodama, F. (1992). Technology Fusion and the New R&D. *Harvard Business Review*: 70–8

Kogut, B. and Zander, U. (1992). Knowledge of the Firm, Combinative Capabilities, and the Replication of Technology. *Organization Science* 3(3): 383–97

Kötter, R. (2002). Magneto Electronics Has Moved. *Chip Unaxis*: 6

Kozin, M. D. and Young, K. C. (1994). Using Acquisitions to Buy and Hone Core Competencies. *Mergers & Acquisitions*: 21–26

Krämer, H. (2001). Unified Platform – A Corporate Vision. Plasma-Therm: Internal Document: St Petersburg

Krug, J. A. and Hegarty, W. H. (1997). Postacquisition Turnover among U.S. Top Management Teams: An Analysis of the Effects of Foreign vs. Domestic Acquisitions of U.S. Targets. *Strategic Management Journal* 18(8): 667–75

Kubicek, H. (1975). *Empirische Organisationsforschung: Konzeption und Methodik* (Empirical Research on Organizations: Concepts and Methods). Stuttgart: Poeschel

Kurr, T. T. (1999). Technologie Due Diligence: Methodik zur strategischen Bewertung von Geschäftskonzepten und deren zugrundeliegenden Produkttechnologien (Technology Due Diligence: Method for the Strategies of Business Concepts and their Related Production Technologies). Diez an der Lahn, Dissertation

Kusewitt, J. B. (1985). An Exploratory Study of Strategic Acquisition Factors Relating to Performance. *Strategic Management Journal* 6: 151–69

Kwak, M. (2002). Shopping for R&D. *MIT Sloan Management Review* 43(2): 9–10

Lane, P. J. and Lubatkin, M. (1998). Relative Absorptive Capacity and Interorganizational Learning. *Strategic Management Journal* 19: 461–77

Lang, H.-C. (1998). *Technology Intelligence: Ihre Gestaltung in Abhängigkeit der Wettbewerbssituation* (Technology Intelligence: Its Design Depending on the Competitive Situation). Zürich: Industrielle Organisation (Industrial Organisetie)

Larsson, R. (1993). Case Survey Methodology: Quantitative Analysis of Patterns Across Case Studies. *Academy of Management Journal* 36(6): 1515–46

Larsson, R. and Finkelstein, S. (1999). Integrating Strategic, Organizational, and Human Resource Perspectives on Mergers and Acquisitions: A Case Survey of Synergy Realization. *Organization Science* 10(1): 89–116

Lee, A. S. (1991). Integrating Positivist and Interpretive Approaches to Organizational Research. *Organization Science* 2(4): 342–65

Leonard-Barton, D. (1992). Core Capabilities and Core Rigidities: A Paradox in Managing New Product Development. *Strategic Management Journal* 13: 111–25

Leontiades, M. (1986). The Rewards of Diversification into Unrelated Business. *Journal of Business Strategy*: 81–7

Levitt, B. and March, J. G. (1990). Chester I. Barnard and the Intelligence of Learning. In Williamson, O. E. (eds.), *Organization Theory*: 11–37. New York: Oxford
Lewellen, G. W. (1971). A Pure Financial Rationale for the Conglomerate Merger. *Journal of Finance* 27: 521–45
Lichtenthaler, E. (2000). *Organisation der Technology Intelligence: eine empirische Untersuchung in technologieintensiven, international tätigen Grossunternehmen* (Organizing Technology Intelligence: An Empirical Study in Technology Intensive International Enterprises). Zürich: Federal Institute of Technology
Link, A. N. (1988). Acquisitions as Sources of Technological Innovation. *Mergers & Acquisitions* 23(3): 36–39
Lintner, J. (1965). The Valuation of Risky Assets and the Selection of Risky Investments in Stock Portfolio and Capital Budgets. *Review of Economics and Statistics* 2: 13–27
Little, A. D. (1986–96). Technology and Innovation Management: Service to Drive High Performance, Growth and Value. www.adl.com
Lubatkin, M. (1983). Mergers and the Performance of the Acquiring Firm. *Academy of Management Review* 8: 218–25
Lubatkin, M., Calori, R., Very, P. and Veiga, J. F. (1998). Managing Mergers Across Borders: A Two-Nation Exploration of a Nationally Bound Administrative Hertiage. *Organization Science* 9(6): 670–84
Lubatkin, M. and O'Neill, H. M. (1987). Merger Strategies and Capital Market Risk. *Academy of Management Journal* 30(4): 665–84
Lucks, K. (2002). Die Organisation von M&A in internationalen Konzernen (The Organization of M&A in International Enterprises). *Die Unternehmung* 56. Jg.(4): 197–211
Lucks, K. and Meckl, R. (2002). *International Mergers & Acquisitions: Der prozessorientierte Ansatz*. Berlin, Heidelberg, New York: Springer-Verlag
Lunnan, R. and Barth, T. (2003). Managing the Exploration vs. Exploitation Dilemma in Transnational 'Bridging Teams'. *Journal of World Business* 38: 110–26
Mace, M. L. and Montgomery, G. G. (1962). *Management Problems of Corporate Acquisitions* (3rd edn). Boston: The President and Fellows of Harvard College
Manne, H. G. (1965). Mergers and the Market for Corporate Control. *Journal of Political Economy* 73–74: 110–120
Marks, M. L. (1982). Merging Human Resources: A Review of Current Research. *Mergers & Acquisitions* 17: 38–44
Marks, M. L. and Mirvis, P. (1985). Merger Syndrome: Stress and Uncertainty. *Mergers & Acquisitions* 17: 50–55
Marks, M. L. and Mirvis, P. H. (1998). *Joining Forces: Making One Plus One Equal Three in Mergers, Acquisitions, and Alliances*. San Francisco: Jossey-Bass
Maxis Management (2000). *Business Due Diligence*. Internal document: St Petersburg
Megantz, R. C. (1996). *How to License Technology*. Chichester: John Wiley & Sons
Mendonca, M. and Kanungo, R. (1994). Managing Human Resources: The Issue of Cultural Fit. *Journal of Management Inquiry* 3(2): 189–205
Mergerstat. (2004). M&A Activity U.S. and U.S. Cross-Border Transactions. *Report*.
Meyer, C. (2001). The Second Generation of Speed. *Harvard Business Review* 79(4)
Mintzberg, H. (1990). Strategy Formation: Schools of Thought. In Frederickson, J. W. (ed.), *Perspectives on Strategic Management*: 105–236. New York: Harper Business
Mintzberg, H. and Lampel, J. (1999). Reflecting on the Strategic Process. *Sloan Management Review*: 21–30
Mitterdorfer-Schaad, D. D. (2001). *Modellierung unternehmensspezifischer Innovations-Prozessmodelle* (Modelling Company-specific Motivation Processes). Zürich: ETH

Mueller, D. C. and Tilton, J. E. (1969). Research and Development Costs as a Barrier to Entry. *Canadian Journal of Economics* 4: 570–9

Müller-Stewens, G. (2000). *Akquisitionen und der Markt für Unternehmenskontrolle: Entwicklungstendenzen und Erfolgsfaktoren* (Acquisitions and the Market for Corporate Control: Trends and Success Factors). Stuttgart: Schäffer-Poeschel Verlag

Muth, J. F. (1961). Rational Expectations and the Theory of Price Movements. *Econometrica* 29: 315–35

Nahavandi, A. and Malekzadeh, A. R. (1988). Acculturation in Mergers and Acquisitions. *Academy of Management Review* 13(1): 79–90

Napier, N. K. (1989). Mergers and Acquisitions, Human Resource Issues and Outcomes: A Review and Suggested Typology. *Journal of Management Studies* 26(3): 271–90

Napier, N. K., Simmons, G. and Stratton, K. (1989). Communication During a Mergers: Experience of Two Banks. *Human Resource Planning* 12: 105–22

Nelson, R. R. and Winter, S. G. (1982). *An Evolutionary Theory of Economic Change.* Cambridge, Mass.: Harvard University Press

Neue Zürcher Zeitung (2002). Phonak hat sick verzettelt. 26.11.02

Neue Zürcher Zeitung (2003). Hightech-Uebernahmen nicht ohne Gefahren (High Tech-Acquisitions not without Risks). New York, 15. Juli, p. 25

Neue Zürcher Zeitung (2004). Bei nüchterner Betrachtung zählt die Leistung (A Critical Look at Performance Counts). Zürich, 19.01.2004

Nonaka, I. (1991). The Knowledge-Creating Company. *Harvard Business Review* November–December: 96–104

Nonaka, I. and Takeuchi, H. (1995). *The Knowledge-Creating Company: How Japanese Companies Create the Dynamics of Innovation.* New York: Oxford University Press

OECD-Economic-Outlook. (2003). Trends in Foreign Direct Investments. Research report 73

Pablo, A. L. (1994). Determinants of Acquisition Integration Level: A Decision-Making Perspective. *Academy of Management Journal* 37(4): 803–86

Penrose, E. T. (1959). *The Theory of the Growth of the Firm.* Oxford: Basil Blackwell

Picot, A., Nordmeyer, A. and Pribilla, P. (2000). *Management von Akquisitionen.* Stuttgart: Schäffer-Poeschel Verlag

Polanyi, M. (1966). *The Tacit Dimension.* New York: Anchor Day Books

Porter, M. E. (1980). *Competitive Strategy: Techniques for Analyzing Industries and Competitors.* New York: The Free Press

Porter, M. E. (1986). *Competition in Global Industries.* Boston, Mass.: Harvard Business School Press

Porter, M. E. (1987). From Competitive Advantage to Corporate Strategy. *Harvard Business Review* 3: 43–59

Porter, M. E. (1996). What Is Strategy? *Harvard Business Review*: 61–78

Prahalad, C. K. and Bettis, R. A. (1986). The Dominant Logic: A New Linkage between Diversity and Performance. *Strategic Management Journal* 7: 485–501

Prahalad, C. K. and Hamel, G. (1990). The Core Competence of the Corporation. *Harvard Business Review*: 79–91

PriceWaterhouseCoopers. (2003). Technology Sector Insights: Analysis and Opinions on M&A Activitiy Report.

Pümpin, C. et al. (1985). Unternehmenskultur – Basis strategischer Profilierung erfolgreicher unternehmen (Corporate Culture – The Basis for Successful Corporations). Die Orientierung, Schriftenreihe der Schweizerischen Volksbank, No. 85

Puranam, P., Singh, H. and Zollo, M. (2003). A Bird in the Hand or Two in the Bush?: Integration Trade-Offs in Technology-Grafting Acquisitions. *European Management Journal* 21(2): 179–84

Ramaswamy, K. (1997). The Performance Impact of Strategic Similarity in Horizontal Mergers: Evidence from the U.S. Banking Industry. *Academy of Management Journal* 40(3): 697–715

Ranft, A. L. and Lord, M. D. (2002). Acquiring New Technologies and Capabilities: A Grounded Model of Acquisition Implementation. *Organization Science* 13 (4 July–August): 420–41

Rappaport, A. (1986). *Creating Shareholder Value*. New York: Free Press

Rappaport, A. (1999). *Shareholder Value*. Stuttgart: Schäffer-Poeschel Verlag

Ravenscraft, D. J. and Scherer, F. M. (1989). The Profitability of Mergers. *International Journal of Industrial Organization* 7: 101–16

Razgaitis, R. (1999). *Early Stage Technologies: Valuation and Pricing*. Chichester: John Wiley & Sons

Reger, G., Blind, K., Cuhls, K. and Kolo, C. (1998). Technology Foresight in Enterprises. Main Results of an International Study by the Fraunhofer Institue for Systems and Innovation Research (ISI) and the Department of R&D Management.

Reißner, S. (1992). *Synergiemanagement und Akquisitionserfolg* (Synergy Management and Acquisition Success). Wiesbaden: Gabler

Ring, P. S. and Ven, A. H. V. D. (1992). Structuring Cooperative Relationships between Organizations. *Strategic Management Journal* 13: 483–98

Rivette, K. G. and Kline, D. (2000). Patent Mapping Your Business Development Strategy, *Rembrandts in the Attic*: 145–71. Boston, Mass.: Harvard Business School Press

Rockholtz, C. (1999a). *Marktwertorientiertes Akquisitionsmanagement: Due Diligence-Konzeption zur Identifikation, Beurteilung und Realisation akquisitionsbedingter Synergiepotentiale* (Market Oriented Acquisition Management: Due Diligence Concept for the Identification, Assessment and Realisation of Acquisition Specific Synergy Potential). Frankfurt am Main: Peter Lang, Europäischer Verlag der Wissenschaften

Rockholtz, C. (1999b). Due Diligence-Konzeption zum synergieorientierten Akquisitionsmanagement (Due Diligence Concept for the Synergy Oriental Acquisition Management). In Berens, W. and Brauner, H. U. (eds.), *Due Diligence bei Unternehmensakquisitionen*. Stuttgart: Schäffer-Poeschel

Roll, R. (1986). The Hubris Hypothesis of Corporate Takeovers. *Journal of Business* 59: 197–216

Rovit, S. (2003). Your Best M&A Strategy. *Harvard Business Review*: 16–17

Rumelt, R. (1974). *Strategy, Structure and Economic Performance*. Boston, Mass.: Division of Research, Harvard University

Sales, A. L. and Mirvis, P. H. (1985). When Cultures Collide: Issues in Acquisition. In Kimberly, J. R. and Quinn., R. E. (eds.), *New Futures: The Challenge of Managing Corporate Transitions*: 107–33. Homewood, Ill.: Dow Jones-Irwin

Salter, M. and Weinhold, W. (1979). *Diversification through Acquisition: Strategies for Creating Economic Value*. New York: The Free Press

Sanchez, R. and Heene, A. (1997). Competence-Based Strategic Management: Concepts and Issues for Theory, Research, and Practice. In Heene, A. and Sanchez, R. (eds.), *Competence-Based Strategic Management*. Chichester: John Wiley & Sons

Sauber, T. (2003). *Innovation Strategy Formulation Processes*. Zürich: Federal Institute of Technology (ETH Zurich)

Sautter, M. T. (1988). *Strategische Analyse von Unternehmensakquisitionen: Entwurf und Bewertung von Akquisitionsstrategien* (Strategic Analysis of Corporate Acquisitions: Design and Assessment of Acquisition Strategies). Frankfurt a.M.

Savioz, P. (2002). *Technology Intelligence in Technology-Based SMEs*. Zürich: Swiss Federal Institute of Technology, Dissertation

Schein, E. H. (1992). *Organizational Culture and Leadership* (2nd edn). San Francisco: Jossey-Bass

Schipper, K. and Tompson, R. (1983). Evidence on the Capitalized Value of Merger Activity for Acquiring Firms. *Journal of Financial Economics* 11: 85–119

Schneider, S. C. and Dunbar, R. L. M. (1987). Takeover Attempts: What Does the Language Tell Us?, Unpublished Manuscript: INSEAD

Schumpeter, J. A. (1934). *Theory of Economic Development*. Cambridge, Mass.: Harvard Business School Press

Schutz, A. (1973). Concept and Theory Formation in the Social Sciences. In Maurich, N. (ed.), *Collected Papers*, vol. 1: 48–66. The Hague: Martinus Nijhoff

Schweiger, D. M. and DeNisi, A. S. (1991). Communication with Employees Following a Merger: A Longitudinal Field Experiment. *Academy of Management Journal* 34: 110–35

Semler, F.-J. (1996). Der Unternehmens- und Beteiligungskaufvertrag (The Acquisition and Participation Contract). In Hölters, W. (ed.), *Handbuch des Unternehmens- und Beteiligungskaufs* (Handbook for the Acquisition of corporations and Assets), vol. 5: 475–565. köln: Schmidt

Sen, F. and Rubenstein, A. H. (1990). An Exploration of Factors Affecting the Integration of In-House R&D with External Technology Acquisition Strategies of a Firm. *IEEE Transactions on Engineering Management* 37(4): 246–58

Seth, A. (1990). Value Creation in Acquisitions: A Reexamination of Performance Issues. *Strategic Management Journal* 11: 99–115

Shanley, M. (1987). *Post Acquisition Management Approaches: An Exploratory Study*. Philadelphia: University of Pennsylvania

Sharpe, W. (1964). Capital Asset Prices: A Theory of Market Equilibrium under Conditions of Risks. *JoF* 3: 425–42

Shelton, L. (1988). Strategic Business Fits and Corporate Acquisition: Empirical Evidence. *Strategic Management Journal* 9: 279–87

Shleifer, A. and Summers, L. H. (1988). Break of Trust in Hostile Takeovers. In Auerbach, A. J. (ed.), *Corporate Takeovers: Causes and Consequences*: 333–56. Chicago: University of Chicago Press

Shrivastava, P. (1986). Postmerger Integration. *The Journal of Business Strategy* 7: 65–76

Sikora, M. (2000). M&A Almanac: 1999 M&A Profile. *Mergers & Acquisitions* 35: 25

Sinetar, M. (1981). Mergers, Morale, and Productivity. *Personnel Journal* 50: 863–67

Singh, H. and Montgomery, C. A. (1987). Corporate Acquisition Strategies and Economic Performance. *Strategic Management Journal* 8: 377–86

Sirower, M. L. (1994). *Acquisition Behavior, Strategic Resource Commitments and the Acquisition Game: A New Perspective on Performance and Risks in Acquiring Firms*. New York: Columbia University

Sirower, M. L. (1997). *The Synergy Trap: How Companies Lose the Acquisition Game*. New York: The Free Press

Slowinski, G., Rafii, Z. E., J.C. Tao, L. Gollob, M. W. Sagal and Krishnamurthy, K. (2002). After the Acquisition: Managing Paranoid People in Schizophrenic Organizations. Research Technology Management, May 01.

Smith, G., Parr, V. and Russel, L. (1994). *Valuation of Intellectual Property and Intangible Assets*. Chichester: John Wiley & Sons

Steingraber, F. G. (2001). The Seven Deadly Sins of Post Merger Integration, Driving Corporate Culture: Aligning your Employees, Your Business Strategy, and Your Brand. Research report, AT Kearney

Steinöcker, R. (1998). *Mergers and Acquisitions: Strategische Planung von Firmenübernamen: Konzeption – Transaktion – Controlling* (Strategic Planning of

Acquisitions: Conception – Transaction – Controlling. Düsseldorf, Regensburg: Metropolitan-Verlag

Sutton, R. I. (1983). Managing Organizational Death. *Human Resource Management* 22: 391–412

Teece, D. J., Pisano, G. and Shuen, A. (1997). Dynamic Capabilities and Strategic Management. *Strategic Management Journal* 18(7): 509–33

Testa, M. R., Mueller, S. L. and Thomas, A. S. (2003). Cultural Fit and Job Satisfaction in a Global Service Environment. *Management International Review* 43(2): 129–48

Tschirky, H. (1998a). Technologie-Management: Schliessung der Lücke zwischen Management-Theorie und Technologie-Realität (Technology Management: Closing the Gap between Management Theory and Technology Reality). In Tschirky, H. and Koruna, S. (eds.), *Technologie-Management: Idee und Praxis*. Zürich: Verlag Industrielle Organisation

Tschirky, H. (1998b). Konzepte und Aufgaben des Integrierten Technologie-Management (Concepts and Tasks of the Integrated Technology Management). In Tschirky, H. and Koruna, S. (eds.), *Technologie-Management: Idee und Praxis*: 194–394. Zürich: Verlag Industrielle Organisation

Tschirky, H. and Koruna, S. (eds.) (1998). *Technologie-Management: Idee und Praxis* (Technology Management: Idea and Practice). Zürich: Orell Füssli Verlag, Verlag Industrielle Organisation

Tschirky, H. (2000). On the Path of Enterprise Science? An Approach to Establishing the Correspondence of Theory and Reality in Technology-inclusive Companies. *International Journal of Technology Management*, vol. 20, no. 314: 405–28

Tschirky, H., Jung, H.-H. and Savioz, P. (2003). *Technology and Innovation Management on the Move*. Zürich: Verlag Industrielle Organisation

Tushman, M. L. and Anderson, P. (1997). *Managing Strategic Innovation and Change*. New York: Oxford University Press

Ulrich, H. (2001). Theoretische versus anwendungsorientierte Wissenschaft (Theoretical versus Applied Research). In Ulrich, H. (ed.), *Gesammelte Schriften*, vol. 5: 211–214. Bern, Stuttgart, Wien: Verlag Paul Haupt

Ulrich, H. and Probst, G. (1988). *Anleitung zum ganzheitlichen Denken und Handeln* (Guide to Holistic Thinking and Acting). Bern: Haupt

Unaxis Management. (2000). Business Due Diligence. Internal document: St Petersburg

Unaxis press release (2001). New COO of Unaxis USA Inc. Appointed, Unaxis press release

Utterback, J. M. (1994). *Mastering the Dynamics of Innovation*. Baston, Mass.: Harvard Business School Press

Utterback, J. M., Allen, T. J., Hollomon, J. H. and Sirbu, M. A. (1976). The Process of Innovation in Five Industries in Europe and Japan. *IEEE Transactions on Engineering Management* 23(2): 3–9

Völker, R. and Kasper, E. (2002). Wertorientierte Technologiebeurteilung (Value Oriented Technology Assessment). *Zeitschrift für Planung* 13: 173–187

Walsh, J. P. (1988). Top Management Turnover Following Mergers and Acquisitions. *Strategic Management Journal* 9: 173–83

Walsh, J. P. (1989). Doing a Deal: Merger and Acquisition Negotiations and their Impact upon Target Company Top Management Turnover. *Strategic Management Journal* 10: 307–22

Walsh, J. P. and Ellwood, J. W. (1991). Mergers, Acquisitions, and the Pruning of Managerial Deadwood. *Strategic Management Journal* 12: 201–17

Weber, R. and Camerer, C. F. (2003). *Cultural Conflict and Merger Failure: An Experimental Approach*. Management Science 49(4): 400–15

Weick, K. E. and Roberts, K. H. (1993). 'Collective Minds in Organizations: Heedful Interrelating on Flight Dects'. *Administrative Science Quarterly*, 38: 357–81

Wernerfelt, B. (1984). A Resource-Based View of the Firm. *Strategic Management Journal* 5: 171–80

Whalley, M. and Heymann, T. (1996). *International Business Acquisitions: Major Legal Issues and Due Diligence*. London: Kluwer Law International

Wolfram, B. (1991). *Strategisches Technologie Management* (Strategic Technology Management), Diss, Wiesbader.

World Law Group, M. F. (1995). *International Business Acquisitions: Major Legal Issues and Due Diligence*. London; The Hague; Boston: Kluwer Law International

Wright, P., Mukherji, A. and Kroll, M. J. (2001). A Reexamination of Agency Theory Assumptions: Extensions and Extrapolations. *Journal of Socio-Economics* 30: 413–29

Yin, R. K. (1994). *Case Study Research: Design and Methods*. Thousand Oaks, London, New Delhi: Sage Publications

Zahn, E. and Weidler, A. (1995). Verwertung technologischer Fähigkeiten – Integriertes Innovationsmanagement (Exploitation of Technological Competencies – Integrated Innovation Management). In Zahn, E. (ed.), *Handbuch Technologie-Management*. Stuttgart: Schäffer-Poeschel

Zahra, S. A. (1999). Technology Strategy and Financial Performance: Examining the Moderating Role of the Firm's Competitive Environment. *Journal of Business Venturing* 11(3): 189–219

Zahra, S. A. and George, G. (2002). Absorptive Capacity: A Review, Reconceptualization, and Extension. *Academy of Management Review* 27(2): 185–203

Zander, U. and Kogut, B. (1995). Knowledge and the Speed of the Transfer and Imitation of Organizational Capabilities: An Empirical Test. *Organization Science* 6(1): 76–92

Zedtwitz, M. v. (2002). Hamba (A): Strategic Fit and Integration of Newly Acquired Firms, *IMD International*. Lausanne, Switzerland: International Institute for Management Development

Index

ABB   230
Absorptive capacity   46, 85, 120, 219
Accelerated momentum   219
Acculturation   193
Acquisition and integration management   138
  four layers of the A&I management   138
  six phases of the A&I management   138
Acquisition and integration process   34, 125
Acquisition capability   113, 119
Acquisition characteristics   105
  business relatedness   106
  market dynamics   109
  relative size   107
  technology development   108
Acquisition experience   34, 119
Acquisition process   126
  holistic acquisition process   127
  integrative acquisition process   126
  systemic acquisition process   127
Acquisition process design   126
Acquisition process tasks   128
Acquisition strategies   105
  play for scale acquisition   105
  substrate for growth acquisition   105
  venturing acquisition   105
Acquisition structure   127, 200
Acquisition type   34, 104
Acquisitions   28
  acquisition types   28
  concentric acquisitions   28
  conglomerate acquisitions   28
  corporate acquisitions   28
  horizontal acquisitions   28
  vertical acquisitions   28
Ammann Lasertechnik   41
Analysis of target's context   170
Analysis of target's resources   169
Aseptic filling   51
Aseptofill   46, 47, 49–57

Assessment of value-creation opportunities   209
Assets   14
  intangible assets   14
  tangible assets   14

Bain & Company   73
Basic technology   109
Batch machine   71
Boundary spanning innovation process   196
Boundary spanning structure   62, 128, 135, 194, 224
Business intelligence   154, 157

Capabilities   14, 15
  combinative capabilities   15
  integration capabilities   15
  organizational capabilities   15
Capital market school   33
Change agents   134
Change management   134
Chief technology officer   127, 198, 219
Ciba Specialty Chemicals   197, 202
Cisco   176, 196, 228, 231
Cluster machine   71
Combinability of technologies   208, 209
Community of practice   157
Company analysis   167
Company screening   154
Competence   14
  core competence   14
  core competence method   24
Competence centre   135, 181, 195, 224
Competence destruction   133
Competitive strategic planning   150, 151
Competitive strategic projects   222
Congruence   117
Consolidation phase   *see* full integration
Contextual difference   46
Contextual fit   85, 113, 123
Contextual match   *see* contextual fit
Contextual similarities   124

Contextual web 215, 216
Corporate coherence 131, 132, 135
Corporate context 123
Corporate control 28
Corporate culture 123
Corporate restructuring phase 132, 133
Cost of capital 14
Creative destruction 239
Cross-selling opportunities 132
Cross-divisional steering committees 195
Cultural clash 130
Cumulated internal frictional resistance 131
CVD technology 71

Deculturation 193
Development strategic planning 150
Development strategic projects 222
Discontinuous market development 108
Discounted cash flow 14
Dominant business logic 123
Dominant design 109
DSP factory 60
Due diligence 202
  environmental due diligence 203
  financial due diligence 202
  legal due diligence 202
  market due diligence 203
  strategic due diligence 203
  tax due diligence 202
  technology due diligence 203
Dynamic technology portfolio 207, 208

Emergence 102
ESEC 69
Evaluation of value-creation opportunities 184
  evaluation of associated risks 187
  evaluation of financial profitability 185
  evaluation of strategic fit 184
  evaluation of timeliness 187
Experimenting knowledge 15
Exploratory research 36, 100
External developments 135

Feasibility check 213
Fillpack 46

Fit 34, 123
  contextual fit 213, 217, 218
  corporate fit 213
  dominant business logic fit 213
  organizational fit 213
Flag methodology 173
Free cash flow 186
Friendly acquisition 104
Full integration 134
Function 20
  functional match 20
Functional analysis 151
Functional layer 141

Gap analysis 151
Gatekeeper 128, 157
General Electric 142, 231

Hearing aid 58
Heckert 86
Hewlett Packard 198, 231
Hillos 43
Hilti 37
Hostile take-over 104
Human integration 130

IBM 144, 201, 228, 231
Identification of innovation opportunities 173, 174
Identification of resource deployment opportunities 178
Information asymmetry 126, 128, 139, 166, 202
Information memorandum 166
Information need 155
Initial conditions 34, 113
Innovation 15, 101
  architectural innovation 17
  incremental innovation 17
  modular innovation 17
  radical innovation 17
Innovation architecture 21, 170, 210, 212
Innovation driven industry 34
Innovation fostering context 124, 125
Innovation management 26
Innovation potential 18
Innovation process 26, 224
Innovation projects 26

Innovation strategy 26
  innovation strategy formulation 26
  innovation strategy implementation 26
Innovation types 17
  business-related innovation 15
  organizational innovation 15
  process innovation 15
  product innovation 15
  technological innovation 15
Innovation-fostering context 216
Integrated management 18
Integrated technology management 17–26
  tasks of the integrated technology management 23
Integration antagonism 130
Integration approach 129, 190, 191
  distant integration approach *see* low integration approach
  full integration approach *see* tight integration approach
  low integration approach 130, 191, 194
  tight integration approach 130, 191
Integration behaviour 130
Integration layer 141
Integration performance measurement 232
Integration planning 220
Integration policy 224, 234
Integration process 129, 131
Integration roadmap 227
Integration steering committee 231
Integration structure 233
Intel 195
Intellectual property 25
  IP licensing 26
  IP management 25
Intellectual property due diligence 206, 207
IP assets 129
IP experts 198

Jenoptik 41

Keep-or-Sell 22
Key technology 109
Knowledge 24
  dispersed knowledge 24
  tacit knowledge 24

Laser distance measurement 42
Leica 41
Letter of intent 202
Leybold 69
Long-term success 101, 102
Loss of innovativeness 107
Low integration phase 132

M&A *see* acquisitions
M&A management 120
Make-or-Buy 22, 151
Management layer 141
Management principles 237
Market pull 21
Market pull analysis 174
Market push analysis 178
McKinsey 49
Mega-merger 35
Mergers *see* acquisitions
Model of reality 100
Model of the acquisition integration process 132
Model of understanding 100
Mutual understanding 134, 135

Need-to-have criteria 155
Net present value (NPV) 117, 185
Nextral company 71
Nice-to-have criteria 155
Non-disclosure agreement 166

Oerlikon Bührle Group (OBH) 69
Operating cash flow 30
Organizational and behavioural school 33
Organizational competence 154
Organizational context 123
Organizational integration 103, 132

Pacemaker technology 108
Pattern of occurrence 102
PECVD 75
Permanent integration teams 228
Phoenix 152
Phonak 57
Photomasking 75
Plasma-Therm 69
Process design 34, 125
Process school 33
Product platform 101, 135

Prospective perspective 240
PVD technology 71

Quick gains 101, 102, 132

Relatedness 34
Relative size 34
Relative standing 201, 239
Research school 33
Resource characteristics 115
  context dependence 115
  dispersion 115
  explicit resources 115
  tacit resources 115
Resource deployment 15, 101
Resource integration 14
  resource fusion 14
  resource leveraging 14
  resource reconfiguration 14, 15
  resource substitution 14
  resource transfer 14
Resource relatedness 117, 118
  complementary resources 118
  redundant resources 118
  supplementary resources 118
  unrelated resources 118
Resources 14
  Context specificity 14
  Knowledge-based resources 14
  Resource dispersion 14
  Resource fungibility 14
  Resource tacitness 14, 15
Retention 103, 131, 133
Retention management 134
Return on investment (ROI) 14, 117
Roadmap 179
Roland Berger 88
Rotating laser 43

S-curve 107
Second wave integration 133, 134
Self due diligence 153
Semiconductor industry 69
Sense of urgency 85, 133, 154
Shareholder value 30
Siemens 229
Slack 135
Socio-cultural integration 222
Socio-technical systems 18
Starrag 86

Starrag-Heckert 86
Steering mechanism 227
Story-telling 131
Strategic business field 20
Strategic fit 113, 121, 122
Strategic options 151
Strategic planning 148
Strategic R&D projects 25
Strategic school 33
Strategic technology decision 22
  Trilogy of strategic technology
    decisions 22
Strategic technology field 18
Strategic technology platform 18, 20
Strategy formulation 138
Strategy formulation process 126
Strategy implementation 138, 228
Strategy processes 125
Supportive integration 132, 222
Supportive layer 141
Symbolic actions 131
Synergies 132

T&I management 120
Task force 52
Task integration 130, 222
Technological mobility 107
Technology 14
  core technology 20
  process technology 14
  product technology 14
Technology assessment 24
Technology attractiveness 206
Technology base 114, 206
  ability to combine the technology
    base 115
  attractiveness of the technology base
    115
  level of mastery of the technology
    base 115
Technology control 24
Technology due diligence 128, 204
  first level technology due diligence
    204–9
  second level technology due diligence
    209–13
  third level technology due diligence
    213–18
Technology intelligence 25, 154, 157
Technology management 17–26

Technology marketing 25
Technology pipeline 209
Technology planning 23
  technology planning process 23
Technology platform analysis 178, 181
Technology platforms 135
Technology portfolio 25, 151
Technology potential 18
Technology push 21
Technology-related risk 128
Technology strategy 18, 24
  technology strategic goal 18
  technology strategic path 21
  technology strategy formulation 24
  technology strategy implementation 24
Technology-based strategic acquisition and integration management 137
Technology-based value creation potential 113, 114
Technology-based value creation 13, 100, 101

Theoretical model 34
Think Tank 91
Time horizon 102
Toccata 60
Transition manager 144, 200, 220
Trust 134

Unaxis 69, 207

Value chain analysis 170, 178
Value creation 14
Value creation opportunity 117, 210
  congruence 210
  feasibility 210
  relatedness 210
  timeliness 210
Value driver 181
Value sustaining 133

Wheel of technology-based value creation 103, 104, 134, 224, 237
Willingness to change 121, 154, 238